INTRODUCTION TO METAL
π-COMPLEX CHEMISTRY

MONOGRAPHS IN INORGANIC CHEMISTRY
Editor: Eugene G. Rochow
Department of Chemistry, Harvard University

INTRODUCTION TO METAL
π-COMPLEX CHEMISTRY

MINORU TSUTSUI
Chemistry Department, Texas A&M University, College Station, Texas

MORRIS N. LEVY
Chemical and Solvent Distillers, Astoria, New York

AKIRA NAKAMURA
Department of Synthetic Chemistry, Osaka University, Toyonaka City, Osaka, Japan

MITSUO ICHIKAWA
Research Laboratories, Japan Synthetic Rubber Company, Ikuta, Mochiizaka, Kawasaki City, Japan

KAN MORI
Basic Research Division, Japan Synthetic Rubber Company, Ikuta, Mochiizaka, Kawasaki City, Japan

ℙ PLENUM PRESS • NEW YORK–LONDON

Library of Congress Catalog Card Number 70-81164

SBN 306-30410-4

© 1970 Plenum Press, New York
A Division of Plenum Publishing Corporation
227 West 17th Street, New York, N.Y. 10011

United Kingdom edition published by Plenum Press, London
A Division of Plenum Publishing Company, Ltd.
Donington House, 30 Norfolk Street, London W.C. 2, England

Foreword

The Plenum Press series, *Monographs in Inorganic Chemistry*, is intended to fill an obvious need for high-level surveys of recent research in that area, particularly in matters which go beyond the traditional or classical boundaries of the subject. The study of π-bonding of hydrocarbon groups (and their derivatives) to metals is exactly that kind of subject, for it provides a new way of understanding the behavior of metals (which constitute four-fifths of all the chemical elements). In addition, π-bonding has expanded the intriguing area of organometallic chemistry threefold, bringing in all the transition metals, the lanthanides, and the actinides. So much has been discovered and developed in the area of π-bonded "complexes" of the metals that important new industrial processes based on such substances have been developed.

A truly comprehensive review of all π-bonded compounds of the metals would now result in an impossibly large and expensive volume, and would require monthly revision. Instead, the present authors have wisely decided to write a survey which outlines the general aspects of preparation, properties, structure, reactions, and uses of such compounds—a survey which can serve as a textbook, but which can also lead the more experienced practitioner to the most advanced literature on the subject. They have clarified and condensed the subject by means of good organization and a liberal use of diagrams—features which will please the general reader. The authors are to be commended for advancing the teaching and understanding of inorganic chemistry, and we are happy to include this book in the series.

E. G. Rochow

Preface

In every generation achievements in science serve mankind. The progress accomplished stimulates the next generation to even greater achievements, which may take the form of increasing, crystallizing, or detailing existing theories. Other forms, generally resulting from persistence and enlightened fortune, open new areas of investigation previously unimagined and have an impact that may be felt for many years.

An example of this latter form of achievement was the preparation and elucidation of the structure of dicyclopentadienyliron (ferrocene), dibenzenechromium iodide, and olefin metal π-complexes, which provided an introduction to a new type of chemical bond, the metal π-complex bond.

The stabilization and isolation of both cyclobutadiene and benzyne derivatives were first achieved by their isolation as metal π-complexes because of academic interest. However, industrial processes such as the Ziegler–Natta olefin polymerization, the Wacker olefin oxidation of alcohols to aldehydes and ketones, and the hydroformylation of olefins, among others, have provided practical applications for achievements in metal π-complex chemistry. Still to be determined are the exact role of metal π-complexes in many vital biological functions and processes such as nitrogen fixation by plants.

Initial progress in the field of metal π-complexes followed the lines of interest generated separately by organic and inorganic chemistry. However, it is becoming increasingly clear that metal π-complex chemistry, bridging both fields, flourishes independently.

As a consequence of the growth of this subject, both the senior and first-year graduate student as well as the working chemist will benefit from this introduction to metal π-complex chemistry. Chapters 1 and 2 provide an introduction to the nomenclature, classification, and preparation of metal π-complexes. Current and historical views on the nature of the bond are presented in Chapter 3 while the implications of the more significant physical properties, structure, and structure determination are discussed in Chapters 4 and 5. Chapter 6 presents a comprehensive study of the reactions of metal π-complexes, although systematic mechanistic treatment awaits future development. An area of continuing interest, catalysis involving metal π-complexes, is discussed in Chapter 7.

Over a 5-year period, we surveyed and classified the literature in order to formalize this new subject of chemistry. We have been fortunate to receive the cooperation of the publishers, and we are grateful to the following chemists who have reviewed the manuscript and offered constructive comments in spite of their busy schedules: M. Brintzinger, J. P. Collman, R. Eisenberg, H. Gysling, M. Hancock, R. F. Heck, R. M. Hedges, D. Lorenz, R. Pettit, M. Rausch, G. Redl, F. J. Smentowski, K. Suzuki, P. M. Treichel, and R. Velapoldi.

We expect to revise the text, particularly with regard to the theoretical and mechanistic treatment of the reactions of metal π-complexes, as new developments occur in this area.

M. Tsutsui
M. N. Levy
A. Nakamura
M. Ichikawa
K. Mori

January 1969

Contents

CHAPTER 1

History, Classification, and Nomenclature

Metal π-complexes possess a relatively new type of direct carbon-to-metal bonding that cannot be designated as one of the classic ionic, σ-, or π-bonds. It is now known that a large number of both molecules and ions such as mono- and diolefins, polyenes, arenes, cyclopentadienyl ions, tropylium ions, and π-allylic ions can form metal π-complexes with transition metal atoms or ions.

These complexes are organometallic compounds by reason of their direct carbon–metal bond, but they also can be described as coordination complexes because the nature and characteristics of the π-ligands are similar to those encountered in coordination complexes.

This chapter reviews the historical development of metal π-complex chemistry and offers an introduction to and examples of the classification and nomenclature encountered throughout the text.

1-1. HISTORY

In 1827 Zeise reported that ethylene reacted with platinum(II) chloride to form a salt $K(C_2H_4)Pt\ Cl_3\ 2H_2O$, but it was not until after the elucidation of the structure of ferrocene in 1952 that attention was redirected to "Zeise's salt," which was proven to be the first reported metal π-complex.

A series of polyphenylchromium compounds that were proposed to be σ-bonded species was reported in 1918 by Hein, who incurred severe criticism because of inconsistencies between the properties exhibited by his poly-phenylchromium compounds and the rather conventional structures proposed for these compounds. The validity of his work was further questioned because only Hein was able to reproduce his results. It was not until 1953 that Hein's syntheses were verified and the polphenylchromium compounds were reformulated as arene π-complexes.

Since the early 1950's, a substantial number of publications have appeared describing new metal π-complexes and elucidating the structures of previously reported species. Generally, these compounds can be classified within three main groups: olefin-, cyclopentadienyl-, and arene-metal π-complexes, leaving mixed complexes to fall within these main groups either by structural or chemical analogies. It should be noted that because of the volume of literature and the significant role of π-allyl complexes in catalysis and polymerization, π-allyl complexes have been treated as a distinct group. However, in the interest of limiting the classification to complex types, π-allyl complexes are logically presented here as olefin-type π-complexes. The remarkable progress in the study of metal π-complexes has contributed not only to the elucidation of the mechanism of known processes such as the Ziegler–Natta polymerization reaction, the oxo reaction, and catalytic hydrogenation but also to the development of the Wacker process for the oxidation of olefins. Within 10–15 years the study of the chemistry of metal π-complexes has rapidly grown into an important and major area in chemistry.

1-2. CLASSIFICATION

The classification of metal π-complexes has customarily been based on the types of organic π-ligands rather than on the metal elements in the π-complex. In accordance with this classification, the three major groups of olefin, π-cyclopentadienyl, and arene metal π-complexes as well as their subgroups are described in the following by illustration of typical π-complexes.

Olefin π-Complexes

Mono-olefins, dienes, polyolefins, and acetylenes are groups that serve as ligands to a transition metal and are classified as olefin π-complexes. Typical examples of olefin π-complexes are

Mono-olefin ligands
Ethyleneplatinum trichloride anion [1-1]*

$$
\left[\begin{array}{c} \text{CH}_2 \\ \| \\ \text{CH}_2 \end{array} \begin{array}{c} \text{Cl} \\ | \\ -\text{Pt}-\text{Cl} \\ | \\ \text{Cl} \end{array} \right]^-
$$

[1-1]

* Double number in brackets refer to structural formulas.

π-Cyclopentadienylcyclopentenerhenium dicarbonyl [1-2]

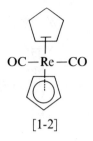

[1-2]

π-En-yl ligands
Bisπ-allylnickel [1-3]

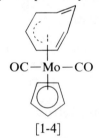

[1-3]

π-Cyclopentadienyl- π-cycloheptatrienylmolybdenum dicarbonyl [1-4]

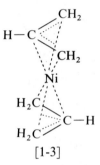

[1-4]

Tropyliumiron tricarbonyl cation [1-5]

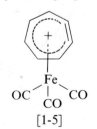

[1-5]

Cycloheptatrienemolybdenum tricarbonyl [1-6]

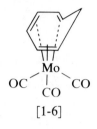

[1-6]

1,4,7-cyclononatrienemolybdenum tricarbonyl [1-7]

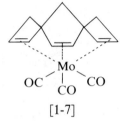

[1-7]

Diene ligands
  Butadieneiron tricarbonyl [1-8]

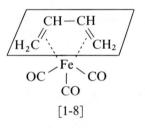

[1-8]

Bis(1,5-cyclooctadiene) nickel [1-9]

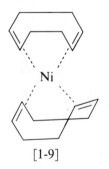

[1-9]

Acetylenic ligands
Diphenylacetylenedicobalt hexacarbonyl [1-10]

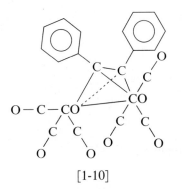

[1-10]

π-Cyclopentadienyl Complexes

A large number of mono-, di-, and tri-π-cyclopentadienyl complexes have been prepared. These π-complexes are classified as follows:

Di-π-cyclopentadienyl complexes (metallocenes)
Di-π-cyclopentadienylmetal [1-11]

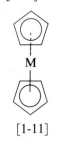

[1-11]

Di-π-cyclopentadienylrhenium hydride [1-12]

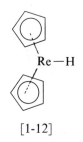

[1-12]

Mono-π-cyclopentadienylmetal nitrosyl and carbonyl complexes
  π-Cyclopentadienylnickel nitrosyl [1-13]

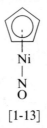

[1-13]

π-Cyclopentadienylmanganese tricarbonyl [1-14]

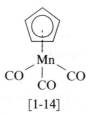

[1-14]

Arene π-Complexes

Complexes in which a benzenoid hydrocarbon serves as a ligand are classified as arene π-complexes.

Diarenemetal π-complexes
  Dibenzene chromium [1-15]

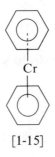

[1-15]

Arenematal carbonyl π-complexes
  Benzenechromium tricarbonyl [1-16]

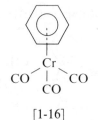

[1-16]

Additional Complexes

Heterocyclic Ring π-Complexes

There are a small number of heterocyclic aromatic ring π-complexes that are classified separately although some aspects of their chemistry parallel either metallocene or arene π-complexes.

Azaferrocene (π-C_4H_4N-π-C_5H_5 Fe) [1-17]

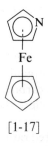

[1-17]

Thiophenechromium tricarbonyl [1-18]

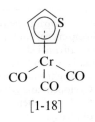

[1-18]

Mixed π-Complexes

Complexes containing more than one type of π-ligand are referred to as mixed π-complexes.

Di-π-cyclopentadienyl allyltitanium(III) [1-19]

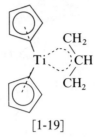

[1-19]

π-Cyclopentadienyl-π-cycloheptatrienyl chromium
π-Cyclopentadienyl-π-cyclobutadiene cobalt.

1-3. NOMENCLATURE

Although this field of chemistry is rapidly developing, no firmly accepted system of nomenclature for metal π-complexes is available. In this text we follow the outline used in *Chemical Abstracts*, which is as follows.

π-Complexes

1. Organic π-ligands precede the metal atom.
2. Organic π-ligands precede inorganic π-ligands.
3. Inorganic π-ligands such as carbonyl or nitrosyls generally follow the metal atom; halides also follow the metal but precede carbonyls or nitrosyls.
4. A prefix such as di- is preferred rather than bis- in describing sandwich-type π-complexes such as dibenzenechromium.

Table 1-1. Nomenclature of π-Complexes

Molecular structure	Nomenclature
$(\pi\text{-}C_5H_5)_2Fe$	Di-π-cyclopentadienyliron (ferrocene)
$(C_6H_6)_2Cr$	Dibenzenechromium
$(\pi\text{-}C_3H_5)_2Ni$	Di-π-allylnickel
$(C_4H_6)Fe(CO)_3$	Butadieneiron tricarbonyl[a]
$(C_6H_6)Cr(CO)_3$	Benzenechromium tricarbonyl
$\pi\text{-}C_5H_5Ni(NO)$	π-Cyclopentadienylnickel nitrosyl
$\pi\text{-}C_5H_5\sigma\text{-}C_5H_5Cr(NO)_2$	π-Cyclopentadienyl σ-cyclopentadienyl-chromium dinitrosyl
$C_{12}H_{18}Ni$	Dodecatriene-centro-nickel

[a] Recently tricarbonylbutadiene-iron(0) is more frequently found in literature, especially in British publications.

5. The symbol π can be used in front of a ligand in order to distinguish π-complex bonding from σ, ionic, or other bonding.

The nomenclature of some typical π-complexes is given in Table 1-1.

Ligands in π-Complexes (Table 1-2)

There are three generally followed rules:
1. The suffix -yl replaces -ium in the accepted ionic or radical name for anionic or radical π-ligands, for example cyclopentadienium becomes π-cyclopentadienyl and allylium becomes simply π-allyl.
2. For cationic π-ligands, the usual ionic suffix, for example, dienium, becomes -dienyl as in cyclohexadienium becoming π-cyclohexadienyl. Cationic ligands sometimes behave as anionic or radical species because of the ligand structure [1-20–1-22].

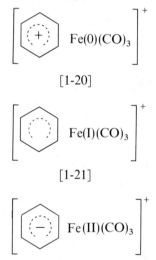

$$[1-20]$$

$$[1-21]$$

$$[1-22]$$

3. In all cases the π-ligand is preceded by a π-notation.

1-4. EXERCISES

1-1. Classify the following compounds:
   a. Styrene palladium chloride. Hint: Pd assumes a square planar structure.

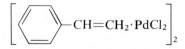

Table 1-2. Nomenclature of Ligands in π-Complexes

Type of ligand	Examples of ligand
Alkene	Ethylene, propylene
π-Allylic	
	π-Allyl (π-C_3H_5) π-Cyclohexenyl (π-C_6H_9)
Diene	1,3-Butadiene
	1,5-Cyclo-octadiene (C_8H_{12})
Dienyl	
	π-Cyclopentadienyl (π-C_5H_5)
Others	
	Dodecatri-2,6,10-ene-1,1-diyl 2,2′-Bi-π-allylene
Dienium	
	π-Cyclohexadienyl ($C_6H_7^+$)
Trienium	
	π-Cyclo-heptatrienyl (tropylium)(π-$C_7H_7^+$)
Acetylene	
	Diphenylacetylene (tolane)
Arene	Benzene, biphenyl, mesitylene

b. Para-divinyl benzenediiron hexacarbonyl. Hint: Fe accepts 10 electrons.

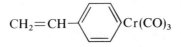

c. Styrene chromiumtricarbonyl Hint: Zero valent chromium.

d. Pyridine chromiumpentacarbonyl. Hint: Cr accepts 12 electrons.

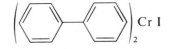

1-2. Give suitable names to the following structures:
   a. $\pi\text{-}C_5H_5Mn(CO)_2 \cdot C_6H_5C \equiv CC_6H_5$
   b. $[(C_2H_4)_2RhCl]_2$
   c. $\pi\text{-}C_5H_5\ Fe(CO)_2 \cdot \sigma\text{-}C_5H_5$
   d. $\pi\text{-}C_5H_5\ Co(CH_3)_2[PPh_3] \longleftarrow [PPh_3]$
   e.

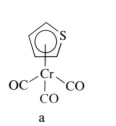

   f. $\pi\text{-}C_5H_5\ Ni\text{-}\pi\text{-}C_3H_5$
   g. $\sigma\text{-}C_3H_5Mn(CO)_5$

1-3. Name the following π-complexes:

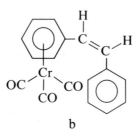
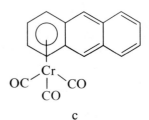

a   b   c

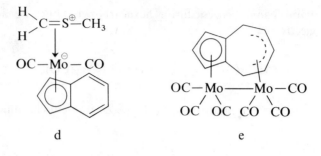

d   e

1-4. Write the structures for the following π-complexes:
a. Cyclobutadieneiron tricarbonyl.
b. Three isomers of $C_8H_8Fe_2(CO)_6$ (C_8H_8 = cyclo-octatetraene).
c. Bicyclo(2.2.1)heptadienemolybdenum tetracarbonyl.

CHAPTER 2

Preparation of Metal
π-Complexes

A large number of metal π-complexes have been prepared since initial work began, and rapid development in this field continues. Among the methods of preparation most frequently used are substitution, elimination, cyclization, ligand or metal exchange, σ–π rearrangements, and redistribution reactions. Examples of these methods are illustrated in this chapter. The equations presented are not necessarily complete, since the emphasis is placed not on a balanced equation but rather on the particular reaction product desired.

2-1. DIRECT SYNTHESES

Many of the earlier syntheses of metal π-complexes involved multistep processes. For example, cyclopentadienyl Grignard was prepared before its reaction with ferric chloride to give ferrocene (2-26).* Although this method is not particularly inconvenient as a laboratory preparation, a more direct one-step synthesis, carried out at 300°C, involves the reaction of cyclopentadiene with hot iron (2-1). This latter method can be used on a commercial scale.

A second one-step synthesis of ferrocene is achieved by the direct action of metal halides on cyclopentadiene in the presence of a base such as di- or triethylamine. This method also can be applied to the preparation of cobalt, nickel, copper, and palladium cyclopentadienyl π-complexes, although the yields with these metals are not as high as with iron.

* Double numbers in parentheses refer to equations.

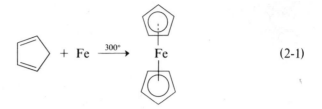

$$\text{+ Fe} \xrightarrow{300°} \quad \text{Fe} \qquad\qquad (2\text{-}1)$$

The original preparation of arene π-complexes is similar to that of ferrocene. Hein's method (see Section 2-8) involved the preparation of an aryl Grignard reagent as the first step of a two-step process. A general one-step method is provided by heating a mixture of a transition metal halide and arene, with a suitable reducing agent such as aluminum powder and aluminum chloride under atmospheric or increased pressure. This is a convenient laboratory method, although the yields sharply decrease Cr > Mo > W (Group VIB) (2-2).

$$3MCl_3 + 2Al + AlCl_3 + 6C_6H_6 \longrightarrow 3 \left[\begin{array}{c} \\ M^+ \\ \\ \end{array} \right] + 3(AlCl_4)^-$$

$$M = Cr, Mo, W \qquad (2\text{-}2)$$

A similar method involves the reaction at elevated temperatures of an arene with a transition metal halide in a hydrocarbon solvent in the presence of a strong reducing agent such as triethylaluminum (2-3). However, this is not strictly a direct synthesis since the preparation of the trialkylaluminum itself requires one step.

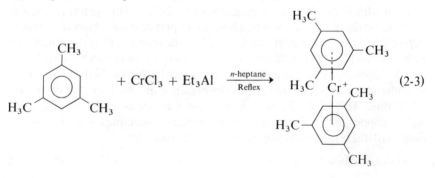

$$+ CrCl_3 + Et_3Al \xrightarrow[\text{Reflex}]{n\text{-heptane}} \qquad\qquad (2\text{-}3)$$

For olefin π-complexes, copper(II) halides as fine powders react with butadiene, norbornadiene, cyclooctadiene or dicyclopentadiene to give the corresponding olefin complex (2-4). Some olefins readily form silver nitrate complexes (2-5) in an aqueous or alcoholic solution.

$$nCuX + olefin \longrightarrow (olefin)(CuX)n \qquad (2\text{-}4)$$

$$mAgNO_3 + nolefin \xrightarrow{\text{in } H_2O} (olefin)n(AgNO_3)m$$

olefin = norbornadiene; 1,3-1,4- and 1,5-cyclooctadiene; cyclooctatetraene

$m = 1,2,3$ (2-5)
$n = 1,2$

2-2. REDISTRIBUTION METHOD

Both arene and cyclopentadienyl π-complexes can undergo redistribution reactions. These reactions involve the exchange of ligands between two species possessing the same or different metal. For example, the reaction of dibenzenechromium with chromium hexacarbonyl provides a method for the preparation of benzenechromium tricarbonyl (2-6). A second example of a

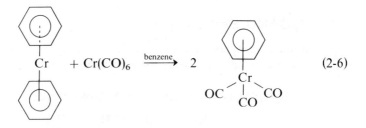

$$(2\text{-}6)$$

redistribution reaction is provided by the preparation of π-benzene π-diphenylchromium(0) (2-7). The redistribution method is not the most commonly used, but it can be useful under limited conditions.

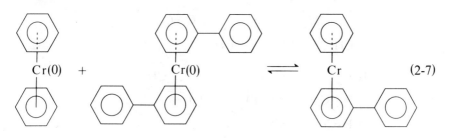

$$(2\text{-}7)$$

2-3. METAL EXCHANGE

Although a metal exchange reaction is not often thought of as a preparative method, such reactions can be useful since they generally proceed in good
yield. Palladium is replaced by iron (2-8) by conversion of tetraphenylcyclobutadiene palladium dibromide to tetraphenylcyclobutadiene iron tricarbonyl. The driving force involves the formation of the more stable tetraphenylcyclobutadiene iron tricarbonyl complex.

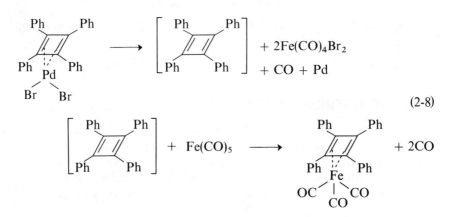

(2-8)

2-4. LIGAND EXCHANGE

Although generally classified under reactions of metal π-complexes,
ligand exchange can also serve as a preparative method (2-9). Similarly,

$$[(C_2H_4)PtCl_2]_2 + 2\ \text{olefin} \longrightarrow [(\text{olefin})PtCl_2]_2 + 2C_2H_4$$

$$\text{olefin} = C_4H_6, C_7H_8, C_{10}H_{16}$$

(2-9)

$$CH_2{=}C{=}CH_2 + (PhCN)_2PdCl_2 \xrightarrow{C_6H_6}$$

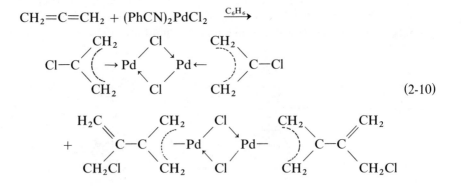

(2-10)

benzonitrile, which is a ligand in dibenzonitrile palladium dichloride $(C_6H_5CN)_2PdCl_2$, can be replaced by a variety of olefins to form olefin π-complexes. Conversely, $PtCl_2(C_6H_5CN)_2$ does not react with olefins; indeed, this compound is formed by the reaction of the olefin complex with benzonitrile.

Di-u-chloro-di-$\pi(\beta$-chloroally)-dipalladium(II) is prepared via the reaction of allene with dibenzonitrile palladium dichloride in benzene (2-10). 1-Methyl- and 1,1-dimethyl-allene also can be used. With benzonitrile as a solvent, only the second product is obtained.

2-5. ADDITION REACTIONS

Addition reactions are often useful in preparing a different complex from an existing one. In many instances the yields are excellent. The addition may

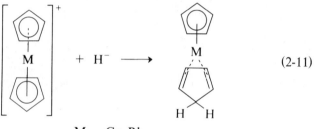

$$M = Co, Rh$$
$$H^- \text{ from } NaBH_4, LiAlH_4$$

occur on the ligand (2-11) and (2-12) or on the metal directly (2-13). Additional examples of both hydrogenation and protonation, which in general may provide the most convenient route for the preparation of a specific complex, are given in Chapter 6.

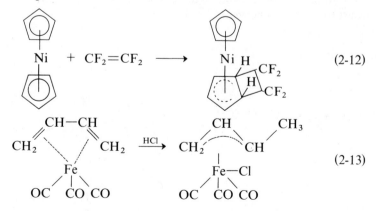

2-6. CYCLIZATION

One of the most novel methods of preparation of metal π-complexes involves the cyclization by trimerization of acetylene and its substituted derivatives. Both arene and olefin π-complexes have been prepared by this method.

As indicated above, manganese and cobalt arene π-complexes could not be prepared by a direct method; however, cyclization of 2-butyne provides a method of preparation of di(hexamethylbenzene) cobalt π-complex (2-14).

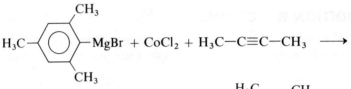

$$(2\text{-}14)$$

A similar method involves the cyclic oligomerization of disubstituted acetylene with other transition metal halides in the presence of a trialkylaluminum in a solvent such as benzene (2-15).

$$CrCl_3 + H_3C\!-\!C\!\equiv\!C\!-\!CH_3 + R_3Al \xrightarrow{\ C_6H_6\ } \qquad (2\text{-}15)$$

Perhaps the most interesting preparative methods are those in which sub-stituted cyclobutadienes are stabilized and isolated as metal π-complexes by the cyclization of diphenylacetylene (2-16)–(2-18).

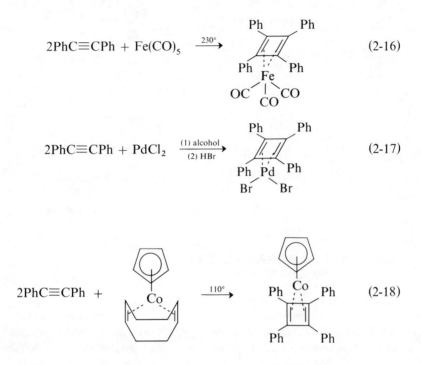

$$2PhC\equiv CPh + Fe(CO)_5 \xrightarrow{230°} \qquad (2\text{-}16)$$

$$2PhC\equiv CPh + PdCl_2 \xrightarrow[(2)\ HBr]{(1)\ alcohol} \qquad (2\text{-}17)$$

$$2PhC\equiv CPh + \qquad \xrightarrow{110°} \qquad (2\text{-}18)$$

Novel mixed π-complexes can be prepared by the reaction of metal halides, metal carbonyls, or their derivatives with substituted acetylenes. As a general method, complexes containing quinone or ketone functional groups can be prepared in this manner (2-19)–(2-21). These reactions should be differentiated from the preceding ones since ultraviolet excitation not only promotes cyclization but also carbonyl insertion.

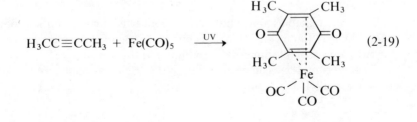

$$H_3CC\equiv CCH_3 + Fe(CO)_5 \xrightarrow{UV} \qquad (2\text{-}19)$$

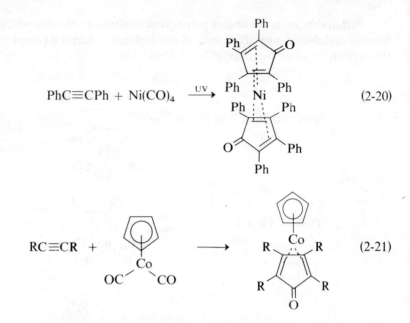

$$PhC\equiv CPh + Ni(CO)_4 \xrightarrow{UV} \qquad\qquad\qquad (2\text{-}20)$$

$$RC\equiv CR + \qquad\qquad\qquad \longrightarrow \qquad\qquad\qquad (2\text{-}21)$$

2-7. σ–π REARRANGEMENTS

In limited cases, σ–π rearrangements (see Section 6-4) can be used effectively as preparative methods.

Diarene π-complexes are prepared by the elimination of stabilizing solvent ligands (for example, THF) on relatively unstable σ-bonded organochromium compounds (2-22). It has been demonstrated that the σ-bonded species

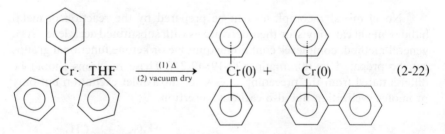

$$(2\text{-}22)$$

need not be isolated as above but may be converted directly to the products within the reaction media (Hein's reaction). Analogous π-complexes can be prepared from isolable dimesitylchromium tetrahydrofuranate.

π-Allyl complexes can be prepared by a σ–π rearrangement of the corresponding σ-allyl compounds under ultraviolet irradiation or by heating (2-23) and (2-24).

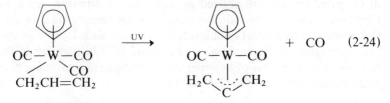

$$(H_2C=CH-CH_2)-Mn(CO)_5 \xrightarrow[\text{or heat}]{\text{UV}} HC\begin{array}{c} CH_2 \\ \diagup \\ \diagdown \\ CH_2 \end{array}-Mn(CO)_4 + CO \quad (2\text{-}23)$$

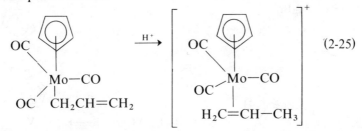

$$(2\text{-}24)$$

σ-Allyl compounds have been converted to π-ethylenic complexes by protonation with acids (2-25). As shown in the reaction, an olefin-cyclopentadienyl complex is formed.

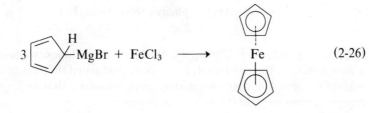

$$(2\text{-}25)$$

2-8. SUBSTITUTION METHODS

As a group, substitution reactions provide the most widely applicable and general methods for the preparation of olefin (including π-allyl), arene, and cyclopentadienyl π-complexes.

The reaction of the appropriate Grignard reagent with selected transition metal halides provides a very useful method for the preparation of many cyclopentadienyl and arene π-complexes. The reaction of cyclopentadienyl-magnesium bromide with ferric chloride constituted the original method for the preparation of ferrocene (2-26).

$$3 \quad \begin{array}{c} H \\ \diagdown \\ \end{array}-MgBr + FeCl_3 \longrightarrow Fe \qquad (2\text{-}26)$$

Because the Grignard reagent acts as a reducing agent, the valence state of the metal is reduced during π-complex formation. Thus, chromium(III) chloride yields di-π-cyclopentadienylchromium, while molybdenum(V) and tungsten(IV) chlorides give $[(\pi\text{-}C_5H_5)_2MoCl]^+$ and $[(\pi\text{-}C_5H_5)_2WCl_2]^+$, respectively. Such a reduction is observed even though equivalent quantities of Grignard reagent are utilized in each case. In addition, the amount of Grignard reagent employed also influences the valence state of the metal. One mole of vanadium(IV) chloride in reaction with 1 mole of cyclopentadienylmagnesium bromide gives a mixture of dihalides, while the same reaction with an excess of cyclopentadienylmagnesium bromide gives di-π-cyclopentadienylvanadium (2-27) and (2-28).

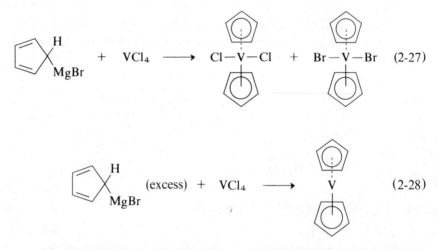

Although the reaction of an aryl Grignard reagent with the metal halide (Hein's reaction) sometimes successfully yields the chromium, molybdenum, and tungsten di-arene π-complexes, it only recommended as a good preparative method for π-arene chromium complexes. Equation (2-29) shows

$$\text{Aryl MgX} + \text{CrCl}n \longrightarrow [(\text{Arene})_2\text{Cr(I)}]^+ \text{ or } [(\text{Arene})\text{Cr(I)}(\text{Ar-Ar})]^+$$
$$+ \text{ others}$$
$$\text{Aryl} = \text{phenyl, tolyl, mesityl} \qquad (2\text{-}29)$$
$$n = 2, 3$$

that, among others, bisdiphenyl-, ditoluene- and dimesitylene-chromium π-complexes are thus prepared. It has been postulated that a σ-bonded aryl-transition metal is initially formed as an intermediate that undergoes a σ–π rearrangement to yield the arene π-complex.)

Many π-cyclopentadienyl complexes may be prepared from the reaction of alkali metal cyclopentadienides with ammonia-soluble transition metal salts such as nitrates and thiocyanates in liquid ammonia (2-30). The amine complexes, $[M(NH_3)_n](C_5H_5)_2$, lose ammonia when heated *in vacuo*, and the uncharged dicyclopentadienyl complexes of iron, cobalt, nickel, chromium, and manganese are obtained.

$$Co(NH_3)_4(SCN)_2 + 2LiC_5H_5 \xrightarrow{\text{liquid } NH_3} [Co(NH_3)_6](C_5H_5)_2 + 2LiSCN$$

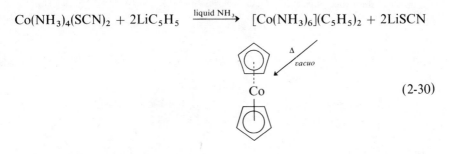

$$(2\text{-}30)$$

Carbonyl π-complexes of all three major groups can be conveniently prepared by the reaction of the appropriate π-system with metal carbonyls. An outline of these methods is given in Table 2-1.

Useful variations of the previous substitution reactions involve the use of mixed metal carbonyls (2-31)–(2-33). In addition, metallocenes at elevated

$$M(CO)nX + olefin \longrightarrow \text{mixed olefin halide } \pi\text{-complex} \quad (2\text{-}31)$$

$$M(CO)nCp + olefin \longrightarrow \text{mixed olefin-}$$
$$\text{cyclopentadienyl } \pi\text{-complex} \quad (2\text{-}32)$$

$$HM(CO)n + diolefin \longrightarrow \text{allyl } \pi\text{-complex} \quad (2\text{-}33)$$

temperatures and under pressure react with carbon monoxide to give π-cyclopentadienyl-metal carbonyls (2-34)–(2-36).

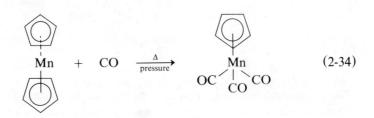

$$(2\text{-}34)$$

Table 2-1. Reactions of π-Systems with Metal Carbonyls

Cyclopentadienyl-(metallocene) type:

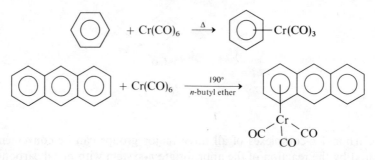

Also useful for dicyclopentadienyl complexes of nickel and iron.

Arene type:

This method is also useful in preparing a number of substituted arene π-complexes that cannot be prepared by other methods:

Olefin type:

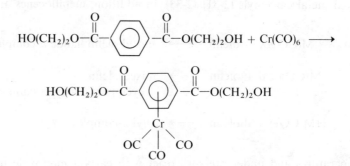

Additional complexes prepared by this method include
(olefin)$_2$Fe(CO)$_3$; (olefin)Fe$_2$(CO)$_6$; (olefin)$_2$Co$_2$(CO)$_4$; (olefin)Co$_2$(CO)$_6$; (diolefin)-M(CO)$_4$, where M = Cr; and Mo and (diolefin)$_2$M(CO)$_2$, where M = Cr, Mo, W.

Table 2-1. Continued

Allyl type:

$$CH_2=CHCH_2Br + NaCo(CO)_4 \longrightarrow$$

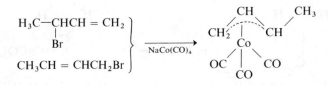

$$+ CO + NaBr$$

Mixtures of 1-bromobutene-2 and 3-bromobutene-1 yield identical 1-methallyl π-complexes:

$$\left.\begin{array}{l} H_3C-CHCH=CH_2 \\ \quad\quad | \\ \quad\quad Br \\[2mm] CH_3CH=CHCH_2Br \end{array}\right\} \xrightarrow{NaCo(CO)_4}$$

Allyl halides with iron monocarbonyl also can produce π-allyl-iron halide carbonyls:

Allyl halide + Fe₂(CO)₉ ⟶

Allyl halide = $CH_2=CHCH_2Br$ $CH_2=CH_2CH-CH_2Cl$ $CH_2=CHCH_2I$

$CH_3CH=CHCH_2Cl$ $CH_3CH=CHCH_2Br$ $CH_2=\overset{\displaystyle |}{\underset{\displaystyle CH_3}{C}}-CH_2Cl$

$CH_3OCOCH=CHCH_2Br$

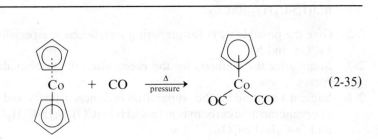

$$(2\text{-}35)$$

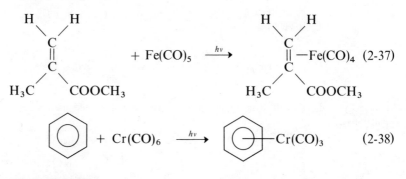

$$\text{Cr} + \text{CO} \xrightarrow[\text{pressure}]{\Delta} [\pi\text{-}C_5H_5Cr(CO)_3]_2 \qquad (2\text{-}36)$$

Irradiation by ultraviolet light has been employed in the preparation of several types of π-complexes, some of which proceed via a substitution process (2-37) and (2-38).

$$+ \text{Fe(CO)}_5 \xrightarrow{hv} \qquad (2\text{-}37)$$

$$+ \text{Cr(CO)}_6 \xrightarrow{hv} \qquad (2\text{-}38)$$

2-9. EXERCISES

2-1. Show the possible preparative route for the following compounds, starting from common metal salts or from metal carbonyls;
   a. $C_4H_6Fe(CO)_3$ (C_4H_6 = 1,3-butadiene)
   b. $(\pi\text{-}C_3H_5)_2Ni$
   c. $(\pi\text{-}C_3H_5)PdCl_2$
   d. $\pi\text{-}C_5H_5Co(CO)_2$
   e. $\pi\text{-}C_5H_5V(CO)_4$
   f. $(CH_3\text{—}C_6H_5)_2V$
   g. $(1,5\text{-}C_8H_{12})_2Ni(C_8H_{12},$ cyclooctadiene)
   h. $[(1,5\text{-}C_8H_{12})RhCl]_2$

2-2. Give the possible ways for preparing metallocenes, especially those of Fe, Co, and Ni.

2-3. Summarize the methods for the preparation of cyclobutadiene complexes.

2-4. Suggest a method for the preparation of dimers, trimers, and tetramers of organometallic carbonyls in $[\pi\text{-}C_5H_5Fe(CO)_2]_2$, $[\pi\text{-}C_5H_5Co(CO)]_3$, and $[\pi\text{-}C_5H_5Fe(CO)]_4$.

2-5. Prepare $(\pi\text{-}C_5H_5)_2Ti(SR)_2$ by usual methods.

2-6. Discuss methods for the preparation of olefin complexes of organometallic compounds.

2-7. Organometallic complexes with a three-electron donating ligand such as N and P have been prepared. Describe the preparation of the Mo and W complexes, for example, $[\pi\text{-}C_5H_5M(CO)_2NO]$, from usual starting materials.

2-8. Various chelate complexes of organometallic complexes have been prepared. Suggest a method for the preparation of $[(\pi\text{-}C_5H_5)_2TiL]^+X^-$, where L = acetylacetone.

2-9. Suggest a method for the preparation of the tris(π-allyl)Ir(III) complex.

2-10. Some organometallic carbonyls undergo carbonyl insertion into a σ-bonded alkyl group. Prepare

$$\pi\text{-}C_5H_5Mo(CO)_2[PPl_3]\cdot\sigma\overset{\displaystyle O}{\overset{\displaystyle \|}{-}C}\text{-}CH_3$$

from usual compounds.

2-10. BIBLIOGRAPHY

G. E. Coates, M. L. H. Green, and K. Wade, "Organometallic Compounds," Vol. 2, Methuen & Co., London (1968).

E. O. Fischer and H. Werner, "Metal π-Complexes," Vol. 1, Elsevier Publishing, Amsterdam, Holland (1966).

P. L. Pauson, "Organometallic Chemistry," St. Martin's Press, New York (1967).

M. Rosenblum, "Chemistry of the Iron Group Metallocenes," Interscience Publishers, New York (1965).

F. G. A. Stone and R. West (eds.), "Advances in Organometallic-Chemistry," Vols. 1–6, Academic Press, New York (1964–1968).

H. Zeiss, P. J. Wheteatley, and H. J. S. Winkler, "Benzenoid-Metal Complexes," Ronald Press, New York (1966).

CHAPTER 3

Nature of the π-Complex Bond

The bonding in transition metal complexes has been described by three different theories: crystal field theory (CFT), valence bond theory (VBT), and molecular orbital theory (MOT). Detailed descriptions of these three approaches are given in the standard inorganic texts and are not repeated here. However, some general statements concerning the applicability of these various bonding descriptions for metal π-complexes are noted.

3-1. CRYSTAL FIELD THEORY (CFT)

The crystal field theory, while having the advantage of being the simplest mathematically, is seriously restricted by its initial premise that the interaction between metal and ligands, resulting in complex formation, is purely electrostatic. Indeed, in cases where the metal–ligand interaction is for the most part electrostatic in nature, CFT does give fairly good agreement with measured spectral and thermodynamic parameters of such species. Complexes containing ligands that are low in the spectrochemical series, especially fluoride complexes, are examples for which such agreement is obtained. The CFT can also give satisfactory results for complexes of nontransition metals. However, two rather apparent failures of the CFT are its inability to explain the great stability of complexes with such Lewis bases as carbon monoxide and trifluorophosphine and the common occurrence of square planar complexes. Simple electrostatics predict only a tetrahedral geometry for a coordination number of four.

29

Thus the CFT formalism, while having the advantages of being relatively simple mathematically and giving an easily understood physical interpretation of the destruction of the fivefold degeneracy of the d orbitals under the influence of a nonspherically symmetrical electric field, clearly has only very limited applicability to the tremendous number of transition metal complexes that have been prepared and well characterized since Werner's time.

3-2. VALENCE BOND THEORY (VBT)

Pauling's valence bond theory is likewise of only limited value in its application to transition metal complexes. In the VBT, the ligand electrons are accommodated in hybrid orbitals localized on the central metal. The orbitals of interest for transition metals are the nd, $(n + 1)s$, $(n + 1)p$, and $(n + 1)d$. An octahedral configuration arises from d^2sp^3 hybridization of the metal orbitals, while dsp^2 hybridization gives the square planar structure and sp^3 hybridization results in a tetrahedral geometry.

Thus one can make a prediction of the geometry of a metal complex based on the number and types of metal orbitals available for bonding. It is expected that the metals at the beginning of a periodic series will have a sufficient number of vacant d orbitals to give octahedral (d^2sp^3) complexes. Also, a

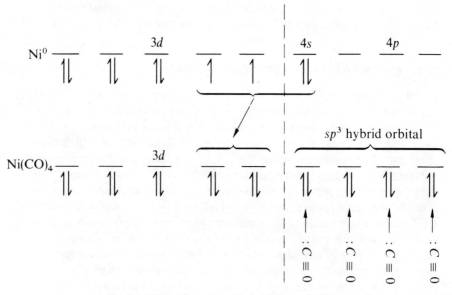

Fig. 3-1. Filling of atomic orbitals of Ni and Ni(CO)$_4$.

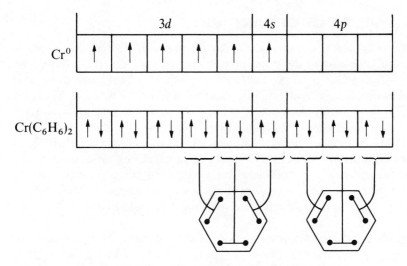

Fig. 3-2. Filling of atomic orbitals of Cr(0) and $Cr(C_6H_6)_2$.

square planar (dsp^2) structure is anticipated for the latter members of a series with a d^7 or a d^8 configuration, having only one vacant d orbital available. However, although complexes of this type are largely square planars as VBT predicts, octahedral and tetrahedral species are also known, emphasizing the limitation of this approach. Finally, in a d^9 configuration, the d orbitals are filled and a tetrahedral (sp^3) complex is anticipated. For example, nickel carbonyl has a tetrahedral configuration, the formation of the required four $4s4p^3$ hybride orbitals on the metal accounting for the experimentally observed geometry. The filling of the nickel atomic orbitals is shown in Fig. 3-1.

Dibenzenechromium(0) has an octahedral configuration consistent with d^2sp^3 hybridization of the chromium atomic orbitals, (Fig. 3-2). If we accept this formulation of the electronic configuration of dibenzene chromium(0), this π-complex can be classified as a coordination compound. The electronic configuration of most metal π-complexes thus can be accounted for by VBT. Therefore, metal π-complexes can be regarded as a new class of coordination compounds.

However, the VBT can be considered as only a limited qualitative explanation of bonding in transition metal complexes. It cannot account for or predict the observed spectra and detailed magnetic properties of transition metal complexes; nor can it predict, even qualitatively, the relative energies of different structures. It nevertheless is quite useful for predicting the geometry of complexes.

3-3. FAILURE OF CFT AND VBT

In recent years, with the spectacular development of physical methods of structural elucidation, the physical bases of the CFT and VBT have been clearly demonstrated to be incorrect. A plethora of evidence from such structural tools apparently indicates that the interaction between metal and ligand is not a purely electrostatic one; rather, it involves the delocalization of electron density over an orbital system involving both metal and ligand. Stated simply, transition metal complexes in general involve covalent bonding between metal and ligands. The degree of covalence varies from a very slight amount to a very considerable amount, depending upon the ligands present in the coordination sphere. Thus it was fortuitous for complexes of the former type to obtain fairly good agreement with observed data by use of CFT.

Probably the most impressive evidence for the existence of such electron delocalization over a molecular-type system is that provided by electron spin resonance spectroscopy (ESR). The fine structure observed in paramagnetic complexes with ligands containing an atom having a nuclear spin (for example, ^{13}C, ^{1}H, ^{35}Cl, and ^{14}N) can be explained only by invoking a delocalization of metal electron density onto the ligands. Detailed examination of such ESR spectra can even lead to quantitative results for spin densities at various nuclei on the ligands.

Other evidence for covalence in transition metal complexes has been obtained from nuclear magnetic resonance spectroscopy (NMR), electronic spectroscopy, nephelauxetic effects, and nuclear quadrupole resonance spectroscopy (NQR) and from the detailed magnetic properties of such complexes. Thus, without giving a detailed description of each of these effects (which is beyond the scope of this book), it can be stated unequivocally that considerable physical evidence shows the existence of some degree of covalence in transition metal complexes.

As will be seen later, the types of ligands one finds in typical organometallic π-complexes are the very ones that give rise to such electron delocalization to the most extreme degree. The bonding description of such π-complexes therefore requires a considerably more sophisticated approach than that of CFT or VBT. The molecular orbital theory is such a description. While considerably more complex mathematically, its basic premise that bonding does not result from the interaction of localized orbitals but rather from the formation, under suitable symmetry conditions, of "molecular orbitals" associated with both metal and ligands allows an explanation of the available physical evidence. The MOT is the most generally applicable approach to chemical bonding presently available.

3-4. MOLECULAR ORBITAL THEORY (MOT)

It has been seen that metal π-complexes are among those that are least satisfactorily described by CFT or VBT. Here the problem of bonding can be treated more completely and quantitatively by MOT or LFT.* The usual approach is to use the linear combination of atomic orbitals (LCAO) method. One assumes that when an electron in a molecule is near one particular nucleus, the molecular wave function is approximately an atomic orbital centered at the nucleus. The molecular orbitals are then formed by simply adding or subtracting the appropriate atomic orbitals. For transition metals, the $3d$, $4s$, and $4p$ orbitals are the atomic orbitals of interest. The ligands may have σ- and π-valence orbitals. Once the appropriate atomic orbitals have been selected for the metal and ligands, the next step is to construct the proper linear combination of valence atomic orbitals for the molecular orbitals. The determination of orbital overlaps that are possible because of inherent symmetry requirements is most readily accomplished by application of the principles of group theory. At this point the procedure becomes somewhat arbitrary in that approximate wave functions must be selected to use in the calculations of the so-called overlap integrals and coulomb integrals. Finally, an arbitrary charge distribution is chosen and the orbital energies and interaction energies are calculated, allowing a solution of the secular equation for the energies and coefficients of the atomic wave functions. Then a new initial charge distribution is repeated until self-consistent values are obtained.

As illustrated in Fig. 3-3, the interaction of two atomic s orbitals gives rise to two molecular orbitals (MO's). An orbital in which the electron density is concentrated between the nuclei is referred to as a bonding MO. The symbol σ (sigma) indicated that the orbital in question has cylindrical symmetry about the internuclear axis. The σ-bonding MO is abbreviated σ^b. The other linear combination is formed by the subtraction of the two s orbitals. The resulting MO has a node in the region between the two nuclei, but it will be localized in the space outside the overlap region. An electron in this orbital is less stable than when in an isolated s atomic orbital, and this LCAO is an antibonding MO. Since this MO was also formed from two s atomic orbitals, it is designated σ^*.

There are two distinct ways for two p orbitals to combine. As shown in Fig. 3-3, addition or subtraction of the p_z atomic orbitals results in the formation of either a bonding or antibonding σ-MO, respectively. The use of the

---

* The ligand field theory (LFT) was originally advanced as a corrected CFT. The LFT relies on the use of molecular orbitals and is often used interchangeably with the MOT, especially in inorganic texts. In describing the bonding in metal π-complexes, we have chosen to introduce the use of molecular orbitals as the MOT.

p_z orbitals as those of the three possible sets of p orbitals that form σ-orbitals is simply the result of defining the z axis of the coordinate system as the internuclear axis. The p_x and p_y orbitals, however, are not symmetrical with respect to rotation about the z axis. An LCAO using the two p_x orbitals gives

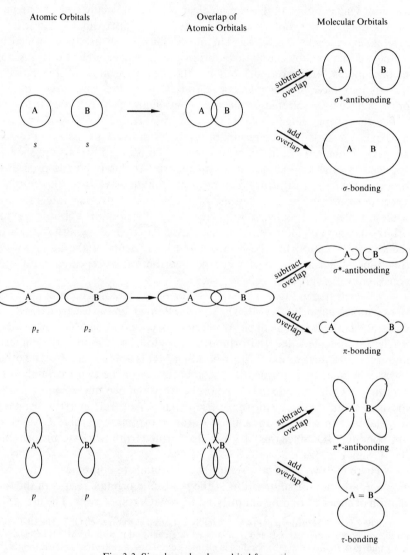

Fig. 3-3. Simple molecular orbital formation.

molecular orbitals having a plus lobe on one side of the z axis and a minus on the other side. Rotation about the z axis by $180°$ (a C_2 operation) simply changes the sign, there being a node in the yz plane. An MO of such symmetry properties is defined as a π-MO. The two p_y atomic orbitals similarly can combine to give π-MO's having nodes in the xz plane. The more stable π^b-MO's will have their electron density concentrated between the nuclei, whereas the less stable π^*-MO's will have nodes between the nuclei. The essential symmetry requirement of a π-MO is that the internuclear axis lies on a nodal plane.

In metal π-complexes, other overlap-utilizing metal d orbitals will arise. Figure 3-4 shows

A. The formation of a σ-type metal–metal bond resulting from d_z^2–d_z^2 overlap.
B. The formation of a metal–ligand bond from the overlap of a metal s and ligand p_z orbital.
C. The formation of a metal–ligand bond from the overlap of a metal p_z with a ligand p_z orbital.
D. The formation of a metal–ligand bond from the overlap of a metal d_z^2 and a ligand p_z orbital.

Thus far we have introduced the molecular orbital concept based on the simple overlap of atomic orbitals. However, this is not the only prerequisite for bond formation. Atomic orbitals are designated not only by their shape

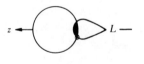

(A) $\sigma(d_{z^2}, d_{z^2})$

(B) $\sigma(s - p_z)$

(C) $\sigma(p_z, p_z)$

(D) $\sigma(d_{z^2}, p_z)$

Fig. 3-4. Examples of σ MO's in metal complexes.

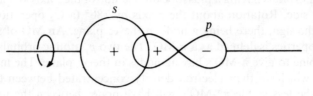

Fig. 3-5. Example of positive overlap.

but also by sign (+ or −). Bond formation is expected only when the inter-
acting orbitals have the same sign (positive overlap; Figs. 3-5 and 3-6).

In this manner, from a knowledge of the available orbitals of various
π-ligands and valence atomic orbitals of the central metal atoms, we can
begin to predict which orbitals will interact.

Figure 3-7 shows the molecular orbitals of the ethylene, π-allyl, and a π-
cyclopentadienyl ligands. The corresponding atomic orbitals of the central
metal atom that can interact are given to the right of each π-ligand.

As an illustrative example, the bonding between the platinum(II) ion and
ethylene in Zeise's salt (Fig. 3-8) involves the following:

1. A σ-bond results from the interaction of the filled π^b-orbital of C_2H_4
   (ψ_1) and a vacant $5d_{x^2-y^2}6s6p_{\overline{xy}}$ hybrid orbital of platinum (Fig. 3-9A).
2. A π-bond results from the overlap of a filled $d_{xy,yz}$ or a hybrid $d_\pi - p_\pi$
   metal orbital with a vacant π*-orbital on the ethylene, ψ_1^* (Fig. 3-9B).

As a consequence of the back donation of metal d electron density into
vacant π*-orbitals on the ethylene, one would predict a decrease in bond
order* of the C—C bond. Such a weakening of the C—C bond is confirmed

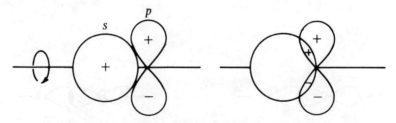

Fig. 3-6. No net overlap; no bond formation expected.

* In MOT, bond order is defined as $\frac{1}{2}$ (number of electrons in bonding orbitals minus the
number of electrons in antibonding orbitals).

π-Cyclopentadienyl

ψ_1
Group theoretical
symbol A_1

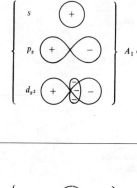

$\begin{Bmatrix} s \\ p_z \\ d_{z^2} \end{Bmatrix}$ A_1 orbital

ψ_2
(E_1)

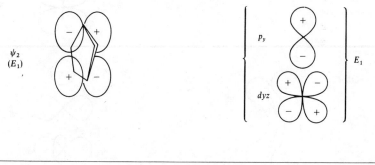

$\begin{Bmatrix} p_y \\ dyz \end{Bmatrix}$ E_1

—————————————————————————————————— (i)

ψ_2'
(E_1)

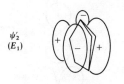

$\begin{Bmatrix} dxz \end{Bmatrix}$ E_1

ψ_3
(E_2)

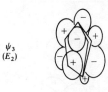 $\cong$

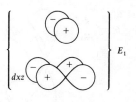

dxy E_2

(continued on p. 38)

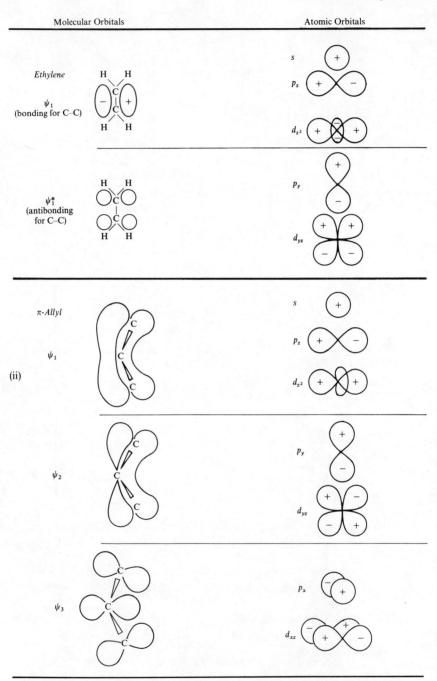

Fig. 3-7. Molecular orbitals of ethylene, π-allyl, and π-cyclopentadienyl ligands with corresponding atomic orbitals of the central metal atom that can interact, possibly to give a π-complex bond.

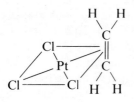

Fig. 3-8. The structure of the ion of Zeise's salt, $(C_2H_4PtCl_3)$. The position of the ethylene ligand, perpendicular to the plane described by the $(PtCl_3)$ moiety, has been established by an x-ray study.

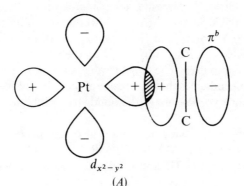

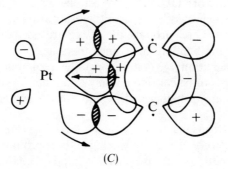

Fig. 3-9. A: σ-Bonding. B: π-Bonding. C: Synergistic effect in $Pt-C_2H_4$ bonding.

by the measured C—C distances in olefin complexes which lie in the range of 1.40–1.47 Å (the normal C=C bond length is ~ 1.34 Å). The reduction in bond order also is reflected in the C—C stretching frequency in the infrared spectra of such complexes (Chapter 4). A displacement to lower wave numbers of this frequency by about $145\,cm^{-1}$ compared to the free corresponding olefins is found in the spectra of numerous platinum–olefin complexes. The C—C stretching frequencies of analogous silver–olefin complexes show a shift of only 50–$70\,cm^{-1}$ toward lower wave numbers, indicating less olefin-to-metal back bonding in these cases—an observation in agreement with the known decreased stability of silver–olefin complexes compared to platinum–olefin complexes. The considerable amount of such infrared data presently available indicates that the extent of the shift of the C—C stretching frequency is a sensitive measure of the bond strength between the olefin and the metal atom.

The phenomenon of back bonding is also responsible for the ability of ligands such as CO, PF_3, NO, RNC, AsR_3, and SbR_3 and other organic π-systems (for example, C_6H_6 and $C_5H_5^-$) to stabilize low oxidation states (Chapter 4). By draining electron density from the central metal, such π-acceptor ligands are able to promote electroneutrality in such compounds. This has been experimentally verified by dipole moment measurements that give unusually low values.

A convenient summary of the relative energy effects can be obtained from the so-called energy level correlation diagram. Let us consider a simple d^6-octahedral complex, ML_6, with no significant π-interaction between metal and ligands. The central metal possesses nine valence orbitals, and the ligands possess six. These can be combined, under the requirements of symmetry, to give 15 molecular orbitals of varying energy (Fig. 3-10). It should be noted that the absolute and even relative energies designated are approximate in nature. The bonding orbitals of the lowest energy are filled first, which is analogous to the expected $d^2_{z^2,x^2-y^2}sp^3$ hybridization in the VBT for an octahedral complex. Three orbitals, the t_{2g} set, are essentially d orbitals and are designated as nonbonding here since there are no available ligand π-orbitals. The designation t_{2g} comes from the terminology of molecular symmetry and needs to be considered only as a label for our purposes. The remaining groups are of higher energy and are designated as antibonding orbitals.

The six d electrons may occupy the t_{2g} level (t_{2g}^6) or both the t_{2g} and e_g^* levels ($t_{2g}^4 e_g^2$). The choice of electron pairing in these orbitals is reflected in the magnetic susceptibility of the complexes (Chapter 4). High-spin complexes result from the latter electron configuration, while low-spin complexes result from the former. The energy difference between the t_{2g} and e_g^* levels

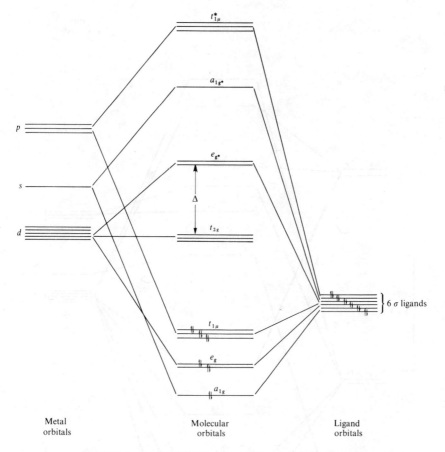

Fig. 3-10. Energy level diagram for a ML_6 octahedral complex.

constitutes the ligand field splitting of the d orbitals. This is similar to the conclusions of the CFT, but here Δ is related to the covalence of the metal–ligand bonding rather than an electrostatic ion–ion or ion–dipole interaction.

In the cases where the ligands are of the π-bonding variety, such as CO or C_2H_4, the previously nonbonding d levels become bonding with respect to the π-orbitals of the acceptor ligand with the resultant stabilization of the t_{2g} level and an increase in Δ. The LFT thus explains the existence of very stable complexes involving rather weak Lewis bases such as CO and PR_3.

Pearson's classification of ligands and metals as soft acids and bases can readily be applied to metal π-complexes. Here the term soft means polariz-able and hard indicates nonpolarizable. According to the general guideline,

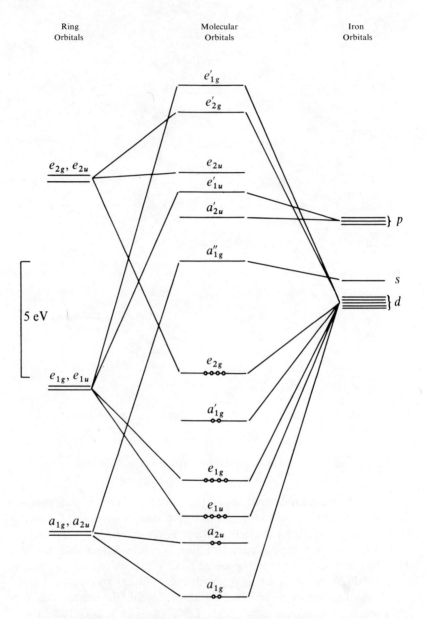

Fig. 3-11. An energy level for ferrocene. The molecular orbital energies are those calculated by Schustorovich and Dyatkina *Doklady. Akad. Nauk. S.S.S.R.* **128**:1234 (1959): using a self-consistent field procedure. The positions of the ring and Fe orbitals on this diagram are only approximate.

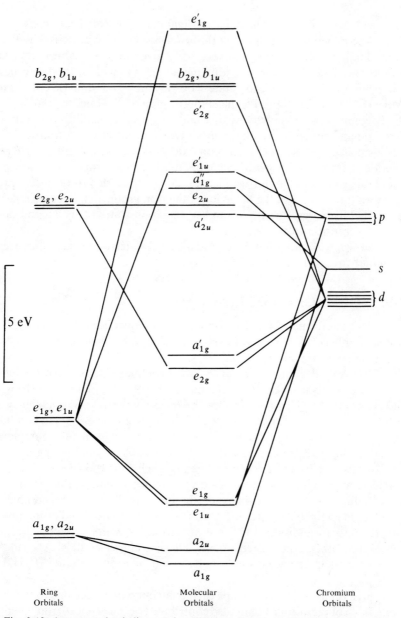

Fig. 3-12. An energy level diagram for dibenzenechromium. The molecular orbital energies are those calculated by Schustorovich and Dyatkina *Doklady. Akad. Nauk. S.S.S.R.* **128**:1234 (1959): using a self-consistent field procedure. The positions of the ring and Cr orbitals on this diagram are only approximate.

hard acids (metals) prefer hard bases (ligands) while soft bases prefer soft acids. The π-ligands considered in this discussion can all be considered soft bases. They therefore would be expected to give the most stable complexes with second- and third-row transition metal ions and with the first transition row elements in low oxidation states. For example, the dimeric complexes $[M(C_2H_4)Cl_2]_2$ (M = Pd, Pt) have been prepared while the analogous Ni(II) complex is unknown. Also, while chemical or anodic oxidation of ferrocene and ruthenocene apparently involve a one-electron transfer process, osmocene loses two electrons on oxidation with silver sulfate or ferric chloride. Under conditions of chronopotentiometric oxidation in acetonitrile, both ruthenocene and osmocene undergo two electron-transfer processes. In this case, also, the behavior of the iron group metallocenes reflects the typical increased stability of higher oxidation states of the second- and third-row transition elements.

The approximate energy level diagrams of ferrocene and dibenzene chromium are given in Figs. 3-11 and 3-12.

3-5. NOBLE GAS FORMALISM

Noble gas formalism or, as it is more commonly referred to, the effective atomic number (EAN) rule is a useful guideline for predicting the stoichoimetries of transition metal π-complexes.

The EAN rule requires that in forming a complex a transition metal tends to accept a number of electrons such that its EAN will equal that of the next highest noble gas. This implies that the metal shows a strong tendency to utilize all nine of its valence orbitals $[nd^5, (n + 1)s,$ and $(n + 1)p^3]$ in bonding. This empirical rule is found to be consistently followed by metal π-complexes as well as those containing ligands that are capable of functioning as both Lewis acids and bases (π-acceptor ligands) such as carbon monoxide, isocyanides, nitric oxide and substituted phosphines, arsines, stilbenes, or sulfides. It fails, however, in many of the classical Werner-type complexes as well as in complexes containing the conjugated dipyridyl and dithioolefin-type ligands. In addition, there are numerous exceptions among the carbonyls and other types listed above. The usefulness of the EAN rule is demonstrated in the very stable cyclobutadieneirontricarbonyl $[C_4H_4Fe(CO)_3]$.

Three donor carbonyl groups each contribute two electrons to the iron atom. Each of the double bonds also contribute two electrons to iron, making a total of ten ligand electrons donated. Zero-valent iron possesses 26 electrons of its own, giving iron a total of 36 electrons. Thus the effective atomic number (EAN) of iron in cyclobutadieneirontricarbonyl equals the atomic

Table 3-1. EAN of the Transition-Metals

Compounds	Electron number	EAN of transition metals
$(C_5H_5)_2Fe$	$26(Fe^0) + 2 \times 5$	36
	$24(Fe^{2+}) + 2 \times 6$	36
$(C_6H_6)_2Cr$	$24(Cr^0) + 2 \times 6$	36
	$26(Fe^0) + 4(\overset{.}{C}H_2-\overset{.}{C}H-\overset{.}{C}H-\overset{.}{C}H_2)$ $+ 3 \times 2(:CO)$	36
	$27(Co^0) + 3H\overset{.}{C}\overset{C-CH_3}{\underset{CH_2}{\big|}} + 3 \times 2(:CO)$	36
	$26(Fe^0) + 5$ $+ 2(:CO)$ $+ 1(\cdot Fe) + 2 \times 1(\cdot CO)$	36
$(\pi\text{-}C_5H_5)NiNO$	$28(Ni^0) + 5$ $+ 3(NO)$	36

number of the next noble gas, krypton. The EAN of the transition metals in the most stable di-π-cyclopentadienyl, diarene, and in olefin π-complexes is often equal to the atomic number of the next noble gas.

The electronic structure of ferrocene may be constructed in either of two ways, both of which obey the EAN rules:

1. Zero-valent iron (26 electrons) and two cyclopentadienyl radicals (10 π-electrons).
2. Divalent iron (24 electrons) and two cyclopentadienide ions (12 π-electrons).

The EAN of iron is 36 in both cases and equals the atomic number of krypton. In a similar way the EAN of the transition metals in $(C_6H_6)_2Cr$, $C_4H_6Fe(CO)_3$, $C_4H_7Co(CO)_3$, and $(C_5H_5)_2Fe_2(CO)_4$ is also 36, as illustrated in Table 3-1. In di-π-cyclopentadienyl di-irontetracarbonyl $(C_5H_5)_2Fe_2$-$(CO)_4$, iron can be considered zero-valent while each cyclopentadienyl ring contributes five electrons.

The compound π-C_5H_5NiNO is regarded as formed by a three-electron donation from the nitric oxide ligand. Most NO complexes are best considered as being formed by an initial one-electron transfer to the metal prior to donation from the nitrosonium ion (NO^+). The nitric oxide ligand, a so-called "odd molecule," is then effectively a three-electron donor. Such a bonding mode is supported by considerable infrared, ESR, and other structural data. It should be noted, however, that recent interpretation of ESR data in terms of MOT indicates that the nitric oxide ligand also can be described by the formal structures $NO\cdot$ and NO^-, in some complexes.

Of course there are a number of exceptions in which a stable structure is obtained without the metal atom having formally attained a noble gas structure. The EAN of planar complexes of Pd(II) and Pt(II) or of the π-allyl complexes of Pd(0) and Ni(0) are less than the atomic number of the next gas. For example, the EAN of palladium in $(C_4H_7)_2Pd_2Cl_2$ (see structure [3-1]) is 52. The atomic number of the next noble gas, xenon, is 54.

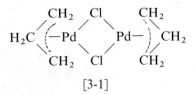

[3-1]

The EAN has been a simple guide for explaining the stability of metal π-complexes and for preliminary characterization of molecular composition.

3-6. EXERCISES

3-1. What is the essential difference between crystal field theory and molecular orbital theory as applied to metal π-complexes?

3-2. The compound $[C_5H_5V(CO)_4]$ can be reduced with sodium in liquid ammonia to give the di-anion $[C_5H_5V(CO)_3]^{2-}$. Which compound should exhibit the lower C—O stretching frequencies in the infrared region?

3-3. The compound $Gd(C_5H_5)_3$ gives an orange compound when treated with a solution of $FeCl_2$. This diamagnetic sublimable compound has a melting point of 174° and shows one absorption in the NMR (5.96τ). Suggest the bonding mode of the cyclopentadienyl group in the rare earth compound.

3-4. Which compound should exhibit the higher C—O stretching frequencies: $(\pi\text{-}C_5H_5)Mo(CO)_3H$ or $(\pi\text{-}C_5H_5Mo(CO)_2NO$?

3-5. For analogous complexes of Co, Rh, and Ir, how will Δ vary as the group is transcended?

3-6. Suggest some other organic species that would give rise to sandwich-type bonding as observed for cyclopentadienyl.

3-7. List the formal oxidation states of various metallocenes in the first row transition elements: Fe, Co, Ni, Ti, V, Mn.

3-8. Assign the EAN for each of the following compounds:

a. $(\pi\text{-}C_5H_5)_2Fe$

b. $[(C_6H_6)_2Cr]I$

c. $(CH_2{=}CH{-}CH{-}CH_2)Fe(CO)_3$

d.

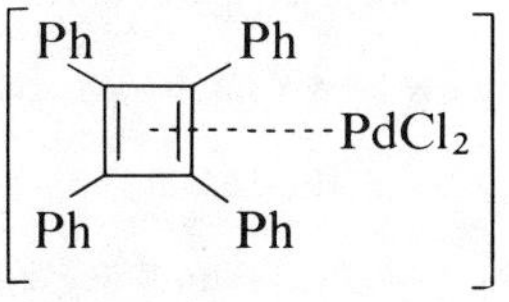

e.

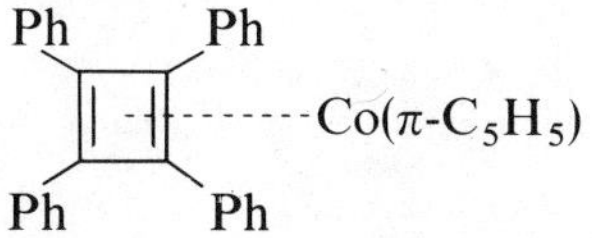

f.

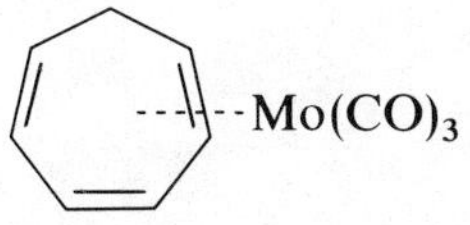

g.

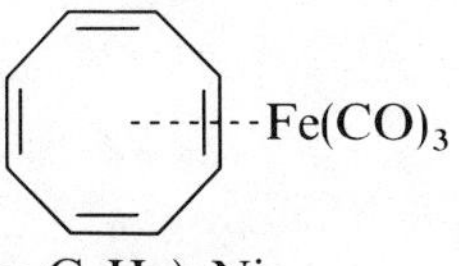

h. $(\pi\text{-}C_3H_5)_2Ni$

i. $(\pi\text{-}C_3H_5PdCl)_2$

j. $\pi\text{-}C_3H_5Co(CO)_3$

3-9.   From a consideration of the EAN rule, what chemical property might V(CO)$_6$ be expected to display?

3-10. The compound [(C$_5$H$_5$)$_2$Mo(NO)I] has recently been reported in the literature. Its infrared spectrum indicates normal NO bonding mode. Suggest a plausible mode of bonding for the cyclopentadienyl groups from a consideration of the EAN rule.

3-7. BIBLIOGRAPHY

C. J. Ballhausen and H. B. Gray, "Molecular Orbital Theory," W. A. Benjamin, New York (1964).

F. Basolo and R. Johnson, "Coordination Chemistry," W. A. Benjamin, New York (1964).

F. Basolo and R. G. Pearson, "Mechanisms of Inorganic Reactions," 2nd ed., Interscience Publishers, New York (1967).

G. E. Coates, "Organometallic Compounds," 2nd ed., Wiley–Methuen, London (1960).

F. A. Cotton, "Chemical Applications of Group Theory," Interscience Publishers, New York (1963)

F. A. Cotton, J. Chem. Educ. 41:466.

F. A. Cotton and G. Wilkinson, "Advanced Inorganic Chemistry," John Wiley & Sons, New York (1966).

F. A. Cotton and G. Wilkinson, "Progress in Inorganic Chemistry," Vol. 1, p. 1, Interscience Publishers, New York (1959).

R. S. Drago, "Physical Methods in Inorganic Chemistry," Reinhold Publishing, New York (1965).

H. B. Gray, "Electrons and Chemical Bonding," W. A. Benjamin, New York (1964).

H. B. Gray, J. Chem. Educ. 41:2

H. H. Jaffe and M. Orchin, "Symmetry in Chemistry," John Wiley & Sons, New York (1965).

S. F. A. Kettle, J. Chem. Educ. 43:21

E. M. Larson, "Transitional Elements," W. A. Benjamin, New York (1964).

A. D. Liehr, J. Chem. Educ. 39:135 (1962).

J. N. Murrel, S. F. A. Kettle, and J. M. Tedder, "Valence Theory," John Wiley & Sons, New York (1965).

L. F. Phillips, "Basic Quantum Chemistry," John Wiley & Sons, New York (1965).

J. M. Richardson, in H. Zeiss (ed.), "Organometallic Chemistry," p. 1, Reinhold Publishing, New York (1960).

CHAPTER 4

Spectroscopic and Magnetic Properties of Metal π-Complexes

Selected physical properties such as spectroscopy and magnetic chemistry reveal useful data on the general skeletal arrangement, bond strength, energy, and valency of metal π-complexes. In this chapter some of the details of infrared spectroscopy (IR), nuclear magnetic resonance (NMR), mass spectra, Mössbauer spectroscopy, magnetic susceptibility, and oxidation state are discussed in terms of the characterizations of metal π-complexes.

4-1. INFRARED SPECTROSCOPY

The normal IR spectral region (4000–650 cm^{-1}) or NaCl region has been the one most generally examined in the spectra of metal π-complexes. In this region one observes the two stretching, wagging (ρ_w), rocking (ρ_r), twisting (ρ_t), and bending (π) modes of the organic group. Recently, with the availability of instrumentation for scanning to longer wavelengths (650–50 cm^{-1}, far infrared region), the bands arising from metal–ligand systems have been measured. This spectral region gives direct information on the strength of the metal–carbon bond. The well-known characteristic positions of bridging and terminal metal–halogen and metal–carbonyl stretching frequencies also have been widely used to deduce the general skeletal arrangement (monomeric vs polymeric) of mixed metal π-complexes.

49

While of definite value in characterization and structural elucidation, IR does not have the wide range of applicability associated with NMR spectroscopy, which is discussed subsequently. The chief reason for this limitation is the complexity of IR spectra and the resulting difficulty in interpretation. However, in relatively simple species, especially those involving olefins and the cyclopentadienyl group, this tool has proven useful in the assignment of bonding modes of the organic moiety. Some of the π-systems subjected to IR studies will now be considered in more detail.

Olefin π-Complexes Ethylene and Mono-olefin (2 π-System)

The IR band most generally considered as the criterion of π-bonding in these systems is the C—C stretching frequency. This band undergoes an appreciable shift (up to 180 cm^{-1}) to lower wave numbers, compared to the free olefin, on complexation (Table 4-1). While IR is inactive in ethylene, the C—C stretching frequency is observed as a weak absorption in the spectra of its complexes. Its bathochromic shift can be attributed to the back donation of metal d electron density into the antibonding orbital of the olefinic system. This results in a reduction of the C—C bond order. An examination of Table 4-1 can reveal some general trends of olefin complexes. On the basis of the lowering of the C—C stretching frequencies, Ag$^+$ forms the weakest ethylene complexes. This is in agreement with other data for olefin complexes of Ag$^+$

Table 4-1. The C—C Stretching Vibrations of Various Mono-olefin Complexes, cm^{-1}

Compound	C—C, cm^{-1}	Magnitude of shift, cm^{-1}
$H_2C{=}CH_2$ (gas)	1623 (Raman)	
$K[Pt(C_2H_4)Cl_3]$	1516	107
$trans[Pt(C_2H_4)(NH_3)Cl_2]$	1521	102
$trans[Pt(C_2H_4)(NH_3)Br_2$	1517	106
$[Pt(C_2H_4)Cl_2]_2$	1516	107
$[Pd(C_2H_4)Cl_2]_2$	1527	96
$K[Pd(C_2H_4)Cl_3]$	1525	98
$[Ag(C_2H_4)]^+$	1550	73
$[(C_2H_4)Mn(CO)_5]^+AlCl_4^-$	1522	101
$[(C_2H_4)Mn(C_5H_5)(CO)_2$	1510	113
$[(C_2H_4)Fe(C_5H_5)(CO)_2]^+$	1527	96
$[(C_2H_4)Mo(C_5H_5)(CO)_3]^+$	1511	112
$[(C_2H_4)W(C_5H_5)(CO)_3]^+$	1510	113
$(C_2H_4)Cr(mesitylene)(CO)_2$	1497	126
$(C_2H_4)Mo(mesitylene)(CO)_2$	1490	133

and, in fact, this property is employed for the purification of olefins. By shaking a wide variety of olefins with aqueous $AgClO_4$, white crystals of stoichiometry $AgClO_4 \cdot n$ olefin ($n = 1$ or 2) are obtained from which the olefin can be readily regenerated. The observed shifts also indicate that the Pt–olefin bond is stronger than the Pd–olefin bond. This is expected on the basis of the increased π-bonding ability of transition metals (in the same oxidation states) as one transcends a periodic group. It is also observed that the two greatest shifts are found for the Cr and Mo mesitylene complexes in which the metals are in a formal zero oxidation state. This is consistent with the general observation that M → L π-bonding increases as oxidation state decreases.

It was recently established that the use of the C—C stretching frequency as a diagnostic criterion of M–olefin bond strength can be misleading in certain cases. This stretching mode can couple with other modes, such as the CH_3 scissoring mode, preventing its general use as a quantitative measure of the strength of the coordinate bond. The metal–olefin stretching force constant has been shown to be the best measure of the strength of coordination. Zeise's dimer and its Pd(II) analogue bands at 408 and 427 cm^{-1}, respectively, have been assigned to the Pt–ethylene and Pd–ethylene stretching modes. The corresponding force constants (in millidynes per angstrom) are Pt–ethylene stretching, 2.25, and Pd–ethylene stretching, 2.14. It should be pointed out that the lighter mass of Pd relative to Pt is responsible for the higher frequency of the Pd–C_2H_4 stretching vibration relative to that of the Pt–C_2H_4 stretching mode.

In Table 4-2 the Pt–olefin stretching frequencies of various mono–olefin complexes are listed. This vibration characteristically appears between 410 and 380 cm^{-1}.

Table 4-2. The Pt–Olefin Stretching Frequencies of Various Mono-olefin Platinum Complexes, cm^{-1}

Complex	v(Metal–olefin)
$K[Pt(C_2H_4)Cl_3] \cdot H_2O$	407
$K[Pt(C_2D_4)Cl_3] \cdot H_2O$	387
$K[Pt(C_2H_4)Br_3] \cdot H_2O$	395
$K[Pt(C_3H_6)Cl_3]$	393
$K[Pt(trans\text{-}C_4H_8)Cl_3]$	387
$K[Pt(cis\text{-}C_4H_8)Cl_3]$	405
$trans[Pt(C_2H_4)(NH_3)Cl_2]$	383
$trans[Pt(C_2H_4)(NH_3)Br_2]$	383

An example of the use of IR in structural elucidation is found in the recently synthesized a series of compounds: [LM(CO)$_4$] (where M = Cr, Mo, and W, and L = 2-allylphenyldiphenylphosphine) [4-1]. The shifts to

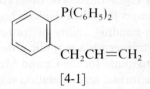

[4-1]

lower wave numbers of the C—C stretching frequencies in these complexes (Table 4-3) support the formulation of this ligand as coordinatively bidentate, bonding through both the phosphorous atom and the olefinic system.

Table 4-3. The C—C Stretching Frequencies of a Series of Chromium Group Metals

Compound	C—C Stretching frequencies, cm^{-1}	Shift, $\Delta\nu$(C=C)
C$_{21}$H$_{19}$P	1637 w*	
[C$_{21}$H$_{19}$PW(CO)$_4$]	1502 vw	140
[C$_{21}$H$_{19}$PCr(CO)$_4$]	1520 vw	122
[C$_{21}$H$_{19}$PMo(CO)$_4$]	1524 vw	118

On the basis of $\Delta\nu$ (C=C), the metal–olefin bond strengths appear to vary in this homologous series in the order W > Cr > Mo. This observation is in agreement with the order of the M—C force constants determined for the Group VI metal hexacarbonyls and also compares with the somewhat greater stability observed for tungsten complexes of simple monoolefins as contrasted with those of the other Group VI metals.

Acetylene π-Complexes

Alkynes, unlike olefins, generally do not react with transition metal complexes to give simple addition products. Rather, the identity of the alkyne is usually lost through a polymerization process, and in the case of carbonyl complexes, CO insertion reactions are common, unsaturated cyclic ketones being among the reaction products. However, in the few cases where

* Double numbers in brackets refer to structural formulas.

the alkyne is incorporated into the complex as such, bathochromic shifts of 110–215 cm^{-1} have been observed. Silver complexes of alkynes generally exhibit C≡C stretching bands at 2070–1930 cm^{-1} (compared to a value of 2230 cm^{-1} for free acetylene). In the series of π-actetylenic complexes of platinous chloride having the general formulas, Na [Pt(RC≡CR′)Cl$_3$] and [(RC≡CR′)PtCl$_2$]$_2$, a downward shift of 189–212 cm^{-1} was observed for the C≡C stretching frequency, a somewhat greater shift than observed for analogous π-olefinic complexes. However, in a series of complexes of the type [Pt(RC≡CR′)(PC$_3$)$_2$] this frequency is lowered to about 1700 cm^{-1}. For example, the value of 1750 cm^{-1} observed for [(PhC≡CPh) Pt(P$_3$)$_2$], almost in the double bond-stretching region, has led to the postulation of the structure in [4-2].

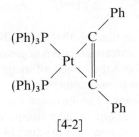

[4-2]

Allyl π-Complexes (3 π-Systems)

In 1960 the nonclassical bonding mode consisting of a three-carbon delocalized allylic system CH$_2$⋯CH⋯CH$_2$, was first proposed. This system, in which all the atoms of the π-allyl group are taken to be coplanar, can be regarded (like NO) as a formal three-electron donor or, if considered in its anionic form, as a four-electron donor. Since this discovery a number of metal π-complexes, originally reported as bis-diene complexes, have been found to actually contain an allylic bonding system. In addition to the parent allyl group, this delocalized (C$_3$)M bonding mode also has been observed in complexes containing the methallyl butadiene, cyclobutenyl, cyclopentaenyl, cyclohexenyl, cyclooctadienyl, and certain pentasubstituted cyclobutadienyl groups. This class of compounds can be considered as intermediate between olefinic and completely delocalized sandwich compounds.

From a study of the IR spectra of a number of allyl compounds of Pd(II) and Ni(II), the most noticeable features observed were the presence of a medium intensity C=C antisymmetric stretching frequency near 1458 cm^{-1}, a position expected for a conjugated double-bond system, as well as a symmetric C=C stretching vibration at 1021, this band being forbidden in ethylene complexes.

An interesting feature of the IR spectra of certain allylic compounds is the presence of an exceptionally low-frequency, strong C—H stretching vibration (about 2850 cm^{-1}). This has been assigned to a geometry in which certain hydrogen atoms (not of the allylic system) are pointed toward the metal (endo) as in the structure [4-3]. Such an anomalous CH stretching vibration also has been observed in other delocalized bonding systems and is generally ascribed to a weakening of the C—H bond resulting from a M—H interaction.

[4-3]

It should be pointed out that the allyl group also can assume a σ-bonding mode. The subject of σ–π-rearrangements is of great interest from a mechanistic point of view and has been the subject of many NMR studies. The σ to π rearrangement of the compound shown in the structure [4-4A] has been followed by IR spectroscopy. The medium intensity band at 1617 cm^{-1} because of the "free" double bond in the σ-compound is replaced by a weaker band at 1560 cm^{-1} in the rearranged product [4-4B], the C—C stretching frequency being lowered by conjugation and coordination to the metal.

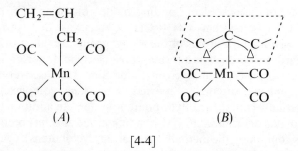

[4-4]

This rearrangement can be rationalized by the increased stability of the π-compound resulting from the loss of a strongly π-bonding CO group from the coordination sphere of the metal—more metal d electron density thereby becoming available for back donation to the allylic π-system.

Butadiene and Cyclobutadiene Complexes (4 π-Systems)

When both double bonds of butadiene are coordinated, there is generally a shift to lower frequencies (Table 4-4).

Table 4-4. $\nu(C=C)$ of 1,3-Butadiene and Its Metal π-Complexes

Compound	$\nu(C=C)$, cm^{-1}
1,3-Butadiene (C_4H_6)	1597
$K_2[(PtCl_3)_2(C_4H_6)]$	1473
$[(\pi\text{-}C_5H_5)Mn(C_4H_6)CO]$	1466
$[(C_4H_6)Fe(CO)_3]$	1464
$[(C_4H_6)Co(CO)_2]_2$	1430

A wide variety of other conjugated diolefins have been shown to form (C_4)M-type bonding systems. These include cyclopentadiene, cyclopenta-dienone, cyclohexadiene, cycloheptatrienone, and cyclooctatetraene. The bands attributed to the free olefin, appearing at 1575–1600 cm^{-1} in the free olefins, display a typical bathochromic shift (to 1500–1460 cm^{-1}) in their complexes. If one double bond of the diolefin is complexed while the second remains "free," often two well-separated absorptions are observed, one being in the region of an uncomplex olefinic linkage and the other at a lower frequency (Table 4-5).

Table 4-5. $\nu(C-C)$ of 1,3-Dienes and Their Metal Complexes (One Double Bond is Uncomplexed)

Compound	$\nu(C=C)$, cm^{-1}
$(\pi\text{-}C_5H_5)Fe(CO)_2$(1,3-butadiene)	1518 and 1626 (free)
$(\pi\text{-}C_5H_5)Fe(CO)_2$(1,3-cyclohexadiene)	1490 and 1621 (free)
1,3-Cyclohexadiene	1600

Cyclobutadiene, an elusive compound long sought by organic chemists, was first stabilized, as the tetramethyl substituted species, in a nickel chloride complex in 1959. Since then a large number of other substituted cyclo-butadiene complexes have been prepared and characterized, culminating in the stabilization of the parent compound in the complex $(C_4H_4)Fe(CO)_3$. Generally, because of the complexity of these molecules, assignments of the IR absorptions are rather difficult and most work in this area has been restricted to comparisons between analogous complexes. For example, most Ph_4C_4 complexes display an intense band of about 1380 cm^{-1}. This band, however, is considerably weakened in intensity when other π-bonding ligands are present.

The tetramethyl-substituted derivatives have proven more susceptible to IR studies. The original nickel chloride derivative has been shown by an x-ray study to be dimeric with two chlorine-bridging atoms and planar four-membered rings centrosymmetrically bonded to the nickel atoms. After assigning the well-known methyl vibrations, those arising from the $(C_4)M$ system and Ni–Cl bonds have been established (Table 4-6).

Table 4-6. Infrared Assignments of $[(CH_3)_4C_4NiCl_2]_2$

cm^{-1}	Assignment
1541 s	C=C (antisym)
1009 w	C=C (sym)
615	
560	CCC deformation of C_4 ring
415	
467	Ni—C str
292 s	Ni—Cl (bridging) str

The use of spectral data, although apparently reasonable and logical, can be misleading. For example, the presence of uncomplexed double bonds in $(C_8H_8)Co(\pi-C_5H_5)$ [4-5] is indicated by an absorption at 1636 cm^{-1} (free cyclooctatetraene absorbs at 1635 and 1609 cm^{-1}). However, there is no corresponding absorption in $(C_8H_8)Fe(CO)_3$. The absorption at 1560 and 1420 cm^{-1} are at lower frequencies than the uncomplexed olefin and strongly indicate that all four double bonds are involved in complex formation. However, x-ray diffraction data proved this view to be incorrect, the structure being as shown [4-6].

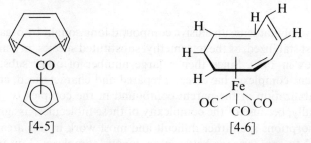

These apparently inconsistent results have been interpreted by postulating a rapid rearrangement from one $(C_4)Fe$ bonding system to the opposite one in solutions of this complex.

π-Cyclopentadienyl Complexes (5 π-System)

The C_5H_5 group can be attached to a metal in four different bonding modes: ionically bonded, centrally σ-bonded, centrally π-bonded, and σ-bonded with a diene structure.

In ionically bonded species, the symmetrical five-member ring is described by D_{5h} symmetry which, from the calculations of group theory, gives rise to four IR active fundamentals (Table 4-7). Some complexes containing the cyclopentadienide group are listed in Table 4-8.

Table 4-7. Characteristic IR Frequencies of Cyclopentadienide Group

1. 3100–3000 cm^{-1}	$v(C-H)$	
2. 1500–1400	$v(C-C)$, or ring deformation	
3. 1010–1000	$\delta(CH)$	
4. 770–660	$\pi(CH)$	

Table 4-8. IR Frequencies of Some Cyclopentadienide Complexes

Complex	(1)	(2)	(3)	(4)
KCp	3048	1455	1009	702
RbCP	3030	1501	1011	696
CsCp	3021	1494	1008	668
CaCp$_2$	3049	1499	1006	760
SrCp$_2$	3077	1435	1008	745
BaCp$_2$	2065	1435	1009	736
EuCp$_2$	3077	1435	1007	739

The local symmetry of the ring is reduced to C_{5v} in centrally bonded complexes with a resultant increase in the number of IR bands from four to seven (Table 4-9). In addition, a metal-ring stretching frequency is present in the far IR in complexes exhibiting this bonding mode. Some typical complexes in this class are listed in Table 4-10.

The IR spectra of the tricyclopentadienyl derivatives of the lanthanides, $Ln(C_5H_5)_3$, exhibits the seven IR active normal vibrations expected for five-membered rings of local symmetry C_{5v}. The CH deformation, γCH, appears at substantially higher frequencies (770–800 cm^{-1}) than in ionic KCp (702 cm^{-1}), indicating considerable reduction of negative charge in the rings and decrease of polarity of M-ring bonding as a result of partial σ-bonding to

Table 4-9. IR Assignments of Centrally σ-Bonded C_5H_5

1. 3100–3020 cm^{-1}	ν(CH)
2. 2970–2900 cm^{-1}	ν(CH)
3. 1450–1410	ν(CC)
4. 1120–1110	ν(CC)
5. 1010–990	δ(CH)
6. 820–740	π(CH)
7. 790–710	π(CH)

Table 4-10. IR Data of Some Centrally σ-Bonded Cyclopentadienyl Complexes, cm^{-1}

Complex	1	2	3	4	5	6	7	ν(M-ring)
LiCp	3048	2906	1426	1120	1003		746	538
NaCp	3048	2907	1422	1144	998		712	315
BeCp$_2$	3067	2941	1433	1121	1008	744	738	414
MgCp$_2$	3067	2913	1428	1108	1004	799	758	439
TlCp	3049	2916	1423	1121	1003	753	734	

the central metal atom. In addition, the spectra of these complexes in the 600–60 cm^{-1} region give an unequivocal indication of covalent M-ring bonding. Intense bands in this region have been assigned to M-ring vibrations. These results are significant in view of the previous ionic formulation of these complexes on the basis of their chemical reactions and magnetic moments, all of which are essentially unchanged from the values of the "free" ions in their simple salts.

Centrally π-bonded complexes represent one of the largest groups of metal π-complexes. The local symmetry here is also C_{5v}, but these compounds are characterized by M-ring vibrations as well as other skeletal modes in the far IR region. In addition, while still exhibiting the seven bands of the centrally σ-bonded cases, the two deformation bands [π(CH)] appear at slightly higher frequencies than observed in the latter cases.

The first IR investigation of such systems involved ferrocene and ruthenocene (Table 4-11). The shift of the C—C stretching frequencies and C—C ring breathing bands to lower frequencies (Table 4-12) indicates weakening of bonding within the ring going from ferrocene to asmocene.

Further support for the greater π-bonding ability of ruthenocene over ferrocene is found in the ring-M-ring symmetric stretching frequency that appears at 332 cm^{-1} in RuCp$_2$, compared to 303 cm^{-1} in Cp$_2$Fe.

Table 4-11. IR Assignments of Some Typical Metallocenes

Compound	ν(CH)	ν(CH)	ν(CC)	ν(CC)	δ(CH)	π(CH)	π(CH)	Antisym ring tilt	R-M str	R-M str force constant
FeCp$_2$	3086	2909	1408	1104	1001	854	814	490	478	2.7
RuCp$_2$	3097		1410	1101	1002	866	821	446	397	2.4
NiCp$_2$	3052	2891	1421	1109	1002	839	772		355	1.5

Table 4-12. Antisymmetric C—C Stretching and C—C Ring-Breathing Bands in Iron Group Metallocenes, cm^{-1}

Compound	Antisym C—C str	Antisym C—C ring breathing
Cp_2Fe	1413	1106
Cp_2Ru	1409	1101
Cp_2Os	1405	1093

A similar phenomenon is observed on comparing a symmetric stretching vibrations of π-bonded metallocenes and ionic cyclopentadienides. The data in Table 4-13 shows that the M-ring bonding decreases in the order Fe > Mg > K.

Table 4-13. Asymmetric C—C Stretching Modes of Typical Cyclopentadienyl Derivatives

Complex	$v(C—C)$, cm^{-1}
$(\pi Cp)_2Fe$	1408
Cp_2Mg	1428
CpK	1443

Absorption frequencies are so characteristic in substituted cyclopentadienyl rings that mono- or di- substitution can readily be detected in products of aromatic substitution reactions that are well known for these "sandwich" compounds.

Arene π-Complexes

The use of IR in the interpretation of structural problems has been well demonstrated for the compound dibenzenechromium. From a comparison of its IR spectrum and the theoretical number of bands for various possible structures, as calculated by the methods of group theory, a sandwich-type structure (D_{6h} symmetry) was proposed. This was subsequently confirmed by an x-ray study. It was also found from a study of the far IR that the M-ring interaction force constant of this compound is slightly lower than that of ferrocene ($2.39 \cdot 10^5$ dynes/cm vs $2.7 \cdot 10^5$ dynes/km).

The characteristic intense infrared absorption bands in these types of complexes can be divided into five frequency ranges: a C—H stretching frequency at 3010–3060 cm^{-1}; a C—C stretching frequency at 1410–1430 cm^{-1};

Table 4-14. Normal Frequencies of
$C_6H_6(D_6A)$

Gaseous phase, cm^{-1}	Liquid phase, cm^{-1}
3073	3062
3064	3053
3057	3048
3056	3048
1599	1594
1482	1479
	1346
1350	1309
1309	1178
1178	1146
1146	1035
1037	1010
1010	993
993	991
990	969
967	
846	850
707	707
693	695
606	606
398	404

a C—C stretching frequency at 1120–1140 cm^{-1}; two or three C—H deformation frequencies at 955–1000 cm^{-1}; and one or two C—H deformation frequencies at 740–790 cm^{-1}. The values given above are also valid for mixed complexes such as $(\pi\text{-}C_5H_5)Cr(C_6H_6)$ and $(C_6H_6)Cr(CO)_3$. For comparison, a complete list of the normal benzene absorption frequencies is given in Table 4-14.

The C_6H_6 vibrations in some di-arene complexes and their cyclopentadienyl and carbonyl derivatives are listed in Tables 4-15 and 4-16.

4-2. NUCLEAR MAGNETIC RESONANCE

Nuclear magnetic resonance spectroscopy of metal π-complexes is widely used currently for identification and characterization of metal π-complexes. Both ^{19}F and ^1H NMR spectra are commonly reported in contemporary literature. With the increasing availability of instruments capable of studying several different isotopes as well as the enhanced resolution of spectra with the development of commercial 100 and 220 Mc instruments,

Table 4-15. Infrared Active Vibrations of π-C_6H_6 in Benzene–Metal Complexes, cm^{-1}

	v(CH)		v(C—C)	σ(C—H)		σ(CH)	
$(C_6H_6)_2Cr$	3037		1426	999	971	833	794
$(C_6H_6)_2Mo$	3030	2916	1425	995	966	811	773
$(C_6H_6)_2W$	3012	2898	1412	985	963	882	798
$(C_6H_6)_2V$, cubic	3058		1416	985	959	818	739
$CpCrC_6H_6$	3061		1410	981	956	838	779
$CpMnC_6H_6$	3049	2916	1427	977		870	808
$[CpCoC_6H_6]PF_6$	3096		1449	986	957	PF_6	

Table 4-16. Infrared Active Vibrations of π-C_6H_6 in Benzene–Metal Carbonyls, cm^{-1}

	v(CH)		v(C—C)	σ(C—H)		σ(C—H)	
$C_6H_6Cr(CO)_3$	3086	2931	1445	1016	978	965	784
$C_6H_6Mo(CO)_3$	3106	2933	1439	1009	1005	967	763
$C_6H_6W(CO)_3$			1433	1003	966	891	776
$[C_6H_6Mn(CO)_3]$ ClO_4	3086		1453 1433	1013		827	
$[(C_6H_6)_2Co_3(CO)_2]^+$	3048		1445	1008	976	908	793
							776

NMR spectroscopy seems certain to become even more useful. In addition to its basic function in locating magnetically active nuclei by the measurement of chemical shift and spin–spin coupling parameters, NMR is also extensively applied today to the study of molecular motions and chemical exchange processes.

The latter experiments are made possible by the inherent small NMR time scale and the experimental ease of carrying out the measurements of NMR signals over a wide temperature range. Variable temperature spectral measurements and spin decoupling experiments have been widely applied to the study of σ–π-rearrangements and "stereochemically nonrigid" molecules. These are molecules that undergo rapid reversible intramolecular rearrangements resulting in geometric equivalence to, but not identical with, the initial configuration. From a detailed analysis of the temperature dependence of the NMR spectrum of an organometallic complex exhibiting this property, it is often possible to establish a mechanism for the rearrangement process. Generally, this phenomenon is accompanied by an increased complexity of the NMR spectrum as the temperature is lowered and a static configuration "freezes out."

In this section a detailed explanation is made of the NMR spectra of some of the more widely studied metal π-complex systems. Chemical shifts are reported tau (τ) values, this parameter being independent of the oscillator frequency used in the measurement and assuming tetramethylsilane (TMS) as the reference compound with $\tau = 10$. Coupling constants (J) are given in cps.

Olefin Complexes

Olefinic protons are shifted from a few tenths to about 3.5 ppm to higher fields on complexation. In an oversimplified view the shift may be regarded as a result of the difference in shielding caused by the metal–olefin bond, as compared with the deshielding due to the uncoordinated π-electrons in the free olefin. The extent of the shift to a higher field is generally reduced, and in some cases a net shift downfield is found in olefins bearing a positive charge. As the metal–proton distance increases, the magnitude of this shift decreases. In addition, proton–proton coupling constants also display changes on co-ordination of an olefin.

Ethylene and Mono-Olefin (2π-System) Complexes

An example of such a complex is Zeise's salt:

$$[C_2H_4PtCl_3]^- \quad \tau 5.3 \, (J_{195_{Pt-H}} = 34 \text{ cps})$$
$$C_2H_4 \qquad \tau 4.0$$

It has been established by x-ray analysis that the metal–olefin axis is perpendicular to the plane of the ethylene molecule. The dipole developed from this metal–olefin bond would point along the same direction as in the ethylene molecule. The magnitude of the shift depends in part upon the extent to which the electrical center of the bond is displaced toward the metal atom. The shift is not exceptionally large—only 1.3τ. The presence of the chlorine may partially cancel the shift to higher field. For example,

$$[(C_2H_4)_2 Re(CO)_4]^+ \quad \tau 6.6$$
$$[(\pi\text{-}C_5H_5)Rh(C_2H_4)_2] \quad \tau 7.25 \text{ (broad) and } \tau 9.0 \text{ (broad)}$$

There are 8 ethylene hydrogens; the reason for the splitting is not clearly understood but probably is the result of [103]Rh–H spin–spin coupling.

Solutions of cationic olefin complexes of silver exhibit NMR spectra that support the conclusion of IR data that the Ag^+–olefin interaction is weak. The olefinic protons here are observed at $\tau \sim 3.9$, about 0.7 ppm lower than the region observed for resonances of the corresponding free olefins. In addition, no spin–spin coupling between $Ag^{107, 109}$ and H was present.

Norbornadiene [4-7] has been shown by NMR to function both as a mono-olefin ligand and a chelating group. In both the free ligand and its metal complexes in which it coordinates as a bidentate ligand there are three different sets of equivalent protons. In complexes in which only one of the double bonds is attached to a metal, five NMR signals are expected [4-7B].

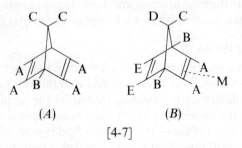

(A)   (B)

[4-7]

In Table 4-17 some NMR data of typical norbornadiene complexes are listed. On the basis of these data $C_7H_8Mn(CO)_2(\pi-C_5H_5)$ is formulated as having a norbornadiene bonding mode. The iron derivative displays an "accidental equivalence" of two types of protons. This also has been observed in other complex organometallic systems and is supported by the relative intensity values.

An example of the value of NMR spectroscopy in establishing rather subtle structures is found in the series of complexes listed in Table 4-18. The NMR data can be interpreted on the basis of the presence of the 2-propenyl-phenyldiphenylphosphine group in the complexes. The double bond of the 2-allylphenyldiphenylphosphine [4-8] that was used in the preparation of the

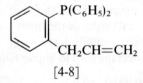

[4-8]

complexes has migrated to the position α to the benzene ring. The NMR spectra of these complexes is entirely consistent with the $-C_1H=C_2HC_3H_3$ structure. The multiplicity of the low field signal in each of the complexes can be explained by invoking $^{31}P-H$ coupling (J = 4.3 cps) across four bonds to explain the further splitting of the doublet because of the olefinic hydrogen adjacent to the phenyl ring. Although this strong spin–spin coupling is not observed for proton–proton interactions, it is quite common for ^{31}P ($I = \frac{1}{2}$) and is a reasonable explanation of the observed double doublet.

Table 4-17. NMR Data for Norbornadiene and Some of Its Complexes

Compound and function	τ	Relative intensity	Multiplicity	J(cps)
C_7H_8, norbornadiene				
$\diagdown$ C—H $\diagup$	3.38	4	3	1.9
$\diagdown$ CH$_2$ $\diagup$	8.05	2	3	1.7
$\diagdown$ C—H $\parallel$	6.50	2	7	1.8
$C_7H_8Cr(CO)_4$				
$\diagdown$ C—H $\diagup$	5.58		3	4.8
$\diagdown$ C—H $\parallel$	6.27		6	
$\diagdown$ CH$_2$	8.70		3	1.4
$C_7H_8Mo(CO)_4$				
$\diagdown$ C—H $\diagup$	5.03		3	2
$\diagdown$ C—H $\parallel$	6.18			
$\diagdown$ CH$_2$ $\diagup$	8.65		3	2
$C_7H_8Fe(CO)_3$				
$\diagdown$ C—H $\diagup$	6.88	6		
$\diagdown$ C—H $\parallel$				
$\diagdown$ CH$_2$ $\diagup$	8.75	2	3	1.4
$C_7H_8Mn(CO)_2(C_5H_5)$				
H$_A$	6.79	2	1	
H$_B$	7.37	2	Complex	
H$_C$	9.91	1	2×3	1.35
H$_D$	9.18	1	2×3	1.65
H$_E$	4.00	2	3	1.75

Table 4-18. NMR Data of $(C_{21}H_{19}P)Cr(CO)_4$ Complexes

Compound	τ^*	Relative intensity	Multiplicity	J (cps) J_{12}	J_{23}
$(C_{21}H_{19}P)Cr(CO)_4$	(1) 4.44	1	2×2		
	(2) 5.42	1	2×4	10.0	6.2
	(3) 8.16	3	2		
$(C_{21}H_{19}P)Mo(CO)_4$	(1) 4.08	1	2×2		
	(2) 5.04	1	2×4	10.5	6.3
	(3) 8.03	3	2		
$(C_{21}H_{19}P)W(CO)_4$	(1) 4.42	1	2×2		
	(2) 5.33	1	2×4	10.0	6.2
	(3) 7.82	3	2		

* Excluding benzene ring protons which have strong absorption of about 2.5τ.

An interesting observation was made in the study of the NMR spectrum of cyclooctatetraeneiron tricarbonyl $(C_8H_8)Fe(CO)_3$. At room temperature a sharp singlet was observed. This strongly suggests that only one type of proton exists or that all the protons are equivalent. Thus a planar structure was postulated for the COT part of the molecule. However, an x-ray crystallographic investigation revealed a nonplanar 1,3-diene bonding mode of the

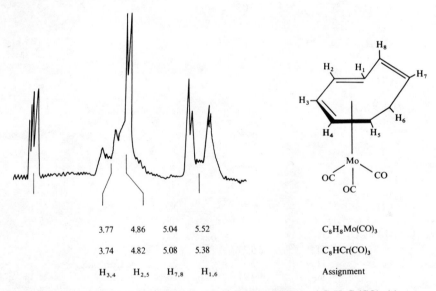

3.77	4.86	5.04	5.52	$C_8H_8Mo(CO)_3$
3.74	4.82	5.08	5.38	$C_8HCr(CO)_3$
$H_{3,4}$	$H_{2,5}$	$H_{7,8}$	$H_{1,6}$	Assignment

Fig. 4-1. Low-temperature NMR spectra of $C_8H_8Mo(CO)_3$ and $C_8H_8Cr(CO)_3$ (τ).

COT. By measuring the NMR spectrum at $-150°$, a more complex spectrum is obtained that is consistent with a 1,3-diene system. These results have been explained by a "ring-whizzing" process resulting from a series of 1,2-shifts involving concomitant migration of double-bond character within the ring. The chromium and molybdenum analogues behave similarly, the "frozen" structures appearing at about $-40°$ for these cases. Both $C_8H_8Mo(CO)_3$ and $C_8H_8Cr(CO)_3$ display one sharp resonance at $+80°$ and multiplet patterns at $-40°$, (Fig. 4-1).

From the spectral changes the rates of rearrangement and free energies of activation were estimated (Table 4-19), giving the following order of increasing difficulty for rearrangement: $C_8H_8Fe(CO)_3 \ll C_8H_8 < C_8H_8Mo$-$(CO)_3 < C_8H_8Cr(CO)_3$.

Table 4-19. Rates of Rearrangement and Free Energies of COT Complexes

Compound	k, sec^{-1}	ΔF^*, kcal/mole
$C_8H_8Mo(CO)_3$	25 (at 10°)	14.8
$C_8H_8Cr(CO)_3$	25 (at 20°)	15.4
$C_8H_8Fe(CO)_3$	200 (at 120°)	7.2

π-Allyl (3 π-System)

The π-allyl radical [4-9] gives three major resonance peaks corresponding to the three types of protons and designated as H_a, H_b, and H_c. The relative intensity ratio is $1:2:2$. H_a ($3 \ll \tau \ll 6$) is the least shielded and under high resolution gives a pattern of overlapping triplets because of the splitting by the two equivalent pairs of protons H_b and H_c. The resonance peaks for H_b(5.6–6.6τ) and H_c(7–8τ) are well separated and appear at generally higher fields.

[4-9]

The NMR data for typical π-allyl complexes are given in Table 4-20.

If the allylic bond is part of a ring, such as in a cyclopentenyl (C_5H_7) system [4-10], H_c is replaced by a methylene group. A resonance peak for the methylene group appears at slightly higher field (8.5–9.2τ).

Table 4-20. NMR Data for π-Allyl (C_3H_5) Complexes

Complex	τ			J_{ac} (cps)	J_{ab} (cps)
	H_a	H_b	H_c		
$[C_3H_5NiBr]_2$	5.7	7.7	8.8	13.3	7.0
$[C_3H_5PdCl]_2$	4.55	5.93	7.00	12.1	6.9
$C_3H_5Co(CO)_3$	5.09	6.94	7.86	10	6
$C_3H_5Mn(CO)_4$	5.30	7.31	8.23	14	10
$C_3H_5NiC_5H_5$	6.30	6.91	8.13	11	3
$C_3H_5PdC_5H_5$	4.32	5.63	6.81	10	5

The study of π to σ rearrangements has been successfully carried out in several allylic systems by variable temperature NMR experiments. The NMR of a $CDCl_3$ solution of allylpalladium(II) chloride (Fig. 4-2A) undergoes a marked change on the addition of 1 mole of DMSO-d_6/mole Pd. The addition of DMSO results in the cleavage of the chlorine bridges, giving the

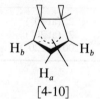

[4-10]

monomeric product shown in Fig. 4-2B. Infrared studies have shown that DMSO coordinates to palladium via the sulfur atom while an oxygen bonding mode is generally found for the first transition row metals.

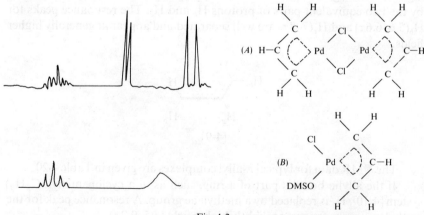

Fig. 4-2.

Temperature, °C

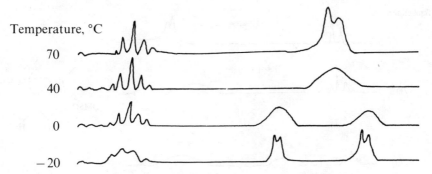

Fig. 4-3. Temperature dependence of NMR spectrum of [DMSO(Cl)Pd(C₃H₅)] in CDCl₃.

The broad profile of the high field signal suggests that some type of re-arrangement or exchange process is operative in the system.

A temperature dependence study of the NMR spectrum of the DMSO adduct (Fig. 4-3) has led to the suggestion that several equilibria are involved in the overall process. A π to σ equilibrium followed by a head-over-tail process (Fig. 4-4) has been postulated to explain the observed spectra. A ligand exchange equilibrium involving the Cl and DMSO-d_6 has also been suggested to explain the equivalence of the two *cis* and two *trans* protons. The other possible explanation for this equivalence is a tetrahedral geometry of the complex. However, the square planar geometry of a great number of analogous Pd complexes is so well established that this explanation is rather unlikely. A ligand exchange equilibrium is further supported by the sensitiv-ity of the system to the concentration of DMSO-d_6 and the observation that there is a shift of $\sim 20°$ in temperature at which the two peaks coalesce in going from a solvent system of pure DMSO-d_6 to one containing 1 mole of DMSO-d_6/mole of Pd in CDCl₃.

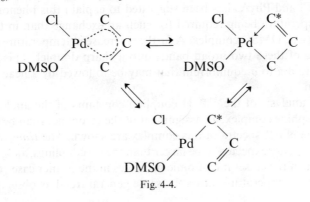

Fig. 4-4.

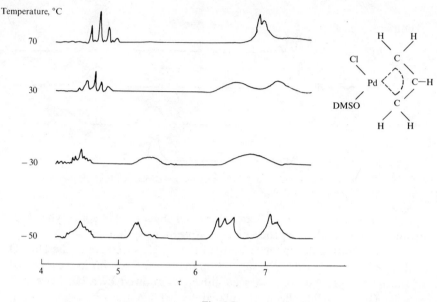

Fig. 4-5.

By preparing the complex containing triphenylarsine in place of DMSO, this exchange process can be demonstrated at low temperatures (Fig. 4-5). This might be expected on the basis of independent IR studies that have shown that triphenylarsine coordinates more strongly to Pd(II) than does DMSO. The presence of five resonances in the $-50°$ NMR supports a square planar configuration for this complex. As the temperature is raised, the resonance of the terminal protons broadens, and at $\sim 30°$ the *cis* protons and the *trans* protons converge into two broad bands. A ligand exchange reaction involving Cl and Ph$_3$As has been suggested to explain this phenomenon, a higher temperature being required for such an exchange than in the previously discussed DMSO complex. A further increase in temperature results in coalescence of these two broad bands into a sharp doublet. This has been attributed to a π to σ equilibrium that may be followed by a head-over-tail equilibrium.

By an analysis of the ^{31}P–H coupling constants of the analogous triphenylphosphine complex, an assignment of the resonances can be made. In Fig. 4-6, the NMR spectra of this complex are shown. The *trans* couplings, $J_{P–H_b}$ or $J_{P–H_c}$, are expected to be larger than the *cis* couplings, $J_{P–P_b}$ or $J_{P–H_c}$, on the basis of the closer interatomic distances in the former case. The effect of raising the temperature shows the same general trend as observed in the

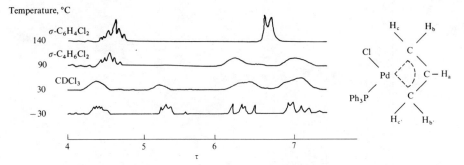

Fig. 4-6. Temperature dependence of NMR spectrum of $[(Ph_3P)ClPd(C_3H_5)]$.

DMSO and Ph_3As analogues. In this case the solvent system must be changed to the higher boiling σ-dichlorobenzene in order to attain a high enough temperature (140°) to observe the well-resolved doublet. The assignments of these three complexes are listed in Table 4-21.

Diolefins (Conjugated or Adjacent Double Bonds; 4 π-System)

As shown in the structure [4-11], a butadiene–metal π-complex also contains three types of protons. The relative resonance intensities, however, are 1:1:1. The resonance signal at 4–5τ is attributed to the H_a protons. The signals for H_b and H_c are again found at a higher field—near 9τ. Several examples of diolefin complexes are given in Table 4-22.

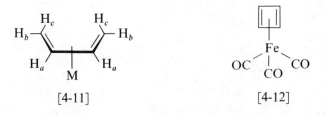

[4-11]   [4-12]

It is interesting to note that cyclobutadieneiron tricarbonyl [4-12], which presumably contains four equivalent protons, gives a single resonance peak at 6.09 τ.

π-Cyclopentadienyl (5 π-System)

The protons in a complexed π-cyclopentadienyl ring system appear to be essentially equivalent. As noted earlier, rapid rotation about the metal-ring

Table 4-21.

Compound	Solvent	T, °C	τ			J, cps
H_c H_b C—CH$_a$ ... Pd ... C H_c H_b / Cl Cl Pd / HC C H H (chloro-bridged dimer)	CDCl$_3$	35	_a_ 4.58	_b_ 5.89	_c_ 6.95	$J_{ab} = 6.8$ $J_{ac} = 11.8$ $J_{bc} = 1.0$
H_c H_b C—CH$_a$ ... Pd ... C H_c H_b / Cl DMSO	CDCl$_3$	−20	4.19	5.71	6.57	$J_{ab} = 6.7$ $J_{ac} = 11.6$ $J_{bc} = 1.0$
	CDCl$_3$	70	_a_ 4.35	_bc_ 6.30		$J_{ab} = J_{ac} = 9.7$
H_c H_b C—CH$_a$ ... Pd ... C $H_{c'}$ $H_{b'}$ / Cl Ph$_3$As	CDCl$_3$	−50	_a_ 4.50	_b_ 5.30 _b'_ 6.75	_c_ 6.35 _c'_ 7.17	$J_{ab} = 6.0$ $J_{ab'} = 6.0$ $J_{ac} = 13.5$ $J_{ac'} = 11.5$

Structure	Solvent	Temp	a	bb′	cc′			Coupling constants
	CDCl₃	30	4.57	6.00	6.73			$J_{ab} = J_{ab'} = 9.7$ $J_{ac} = J_{ac'} = 9.7$
	CDCl₃	70°		bb′cc′				
			4.59	6.41				

			a	b	b′	c	c′	
	CDCl₃	−30	4.42	5.31	6.93	6.30	7.18	$J_{ab} = 7.0$ $J_{ab'} = 6.0$ $J_{ac} = 13.5$ $J_{ac'} = 12.0$ $J_{bP} = 7.0$ $J_{cP} = 10.0$

			a	bb′	cc′	
	σC₆H₄Cl₂	90	4.67	6.11	7.10	
	σC₆H₄Cl₂	140	a	bb′cc′		$J_{ab} = J_{ab'} = —$ $J_{ac} = J_{ac'} = 9.4$
			4.66	6.63		

Table 4-22. NMR Data for Diolefin (Conjugated or Adjacent Double Bond System) Metal π-Complexes

Compound	J	Relative intensity	Multiplicity	J, cps
C_4H_6, butadiene				
$\diagdown$CH$\diagup$	3.8	2	c	
$=CH_2$	4.89	4	3	
$C_4H_6 \cdot Fe(CO)_3$				
$\quad H_a$	4.72	2	c	6.9
$\quad H_b$	8.32	2	2	4–1, 2.5
$\quad H_c$	9.72	2	$2X2$	4–1, 2.1
C_5H_6-Cyclopentadiene				
$\diagdown$CH$\diagup$	3.62	4	c	<1
$\diagdown$CH$_2$$\diagup$	7.21	2	5	~ 1.2
$C_5H_6CoC_5H_5$				
$\quad C_5H_5$	5.41	5	1	
$\quad H_a$	4.80	2	3	~ 2
$\diagdown$CH$_2$$\diagup$	7.32	1	2	12.7
	7.99	1	2	12.7
$\quad H_b$	7.57	2	5	
C_6H_8 Cyclohexa-1,3-diene				
$\quad CH$	4.22	4	2	2
$\quad CH_2$	7.92	4		4
$C_6H_8Fe(CO)_3$				
$\quad H_a$	4.77	2	A_2X_2	$J_{aa} = 4.1$
				$J_{ab} = 6.6$
$\quad H_b$	6.86	2	$2X3$	$J_{ab} = 1.5$
$\quad CH_2$	8.37	4		

axis has been postulated. The resonance signal is generally observed around $5J$. Data for some typical π-cyclopentadienyl complexes are given in Table 4-23.

The compound π-$(C_5H_5)Fe(CO)_2(\sigma$-$C_5H_5)$ has been studied by variable temperature NMR experiments and found to give rather interesting results. An x-ray crystallographic study of this compound has shown that in the solid state one cyclopentadienyl group is π-bonded and the second group is present as a normal σ-bonded 2,4-cyclopentadienyl group. At 30´ the NMR

Table 4-23. NMR Data for Some π-Cyclopentadienyl Complexes

Compound	C_5H_5	H, τ	Miscellaneous, τ
$(\pi\text{-}C_5H_5)_2Fe$	5.96		
$(\pi\text{-}C_5H_5)_2Ru$	5.96		
$(C_5H_5)Mn(CO)_3$	5.8		
$(\pi\text{-}C_5H_5)NiNO$	4.9		
$(\pi\text{-}C_5H_5)Fe(CO)_2CH_3$	5.7		$-CH_3\,9.9$
$[(\pi\text{-}C_5H_5)Cr(NO)_2]_2$	4.82		
$(\pi\text{-}C_5H_5)W(CO)_3H$	5.1	17.4	
$(\pi\text{-}C_5H_5)_2ReH$	6	23.2	
$(\pi\text{-}C_5H_5)_2TiMe$	4.2		

spectrum consists of two resonances, one at $\tau5.60$ and another of essentially the same relative intensity, though broader, at $\tau4.30$. The sharp signal at $\tau5.60$ can be assigned to the $\pi\text{-}C_5H_5$ protons. On decreasing the temperature this signal remains unchanged while the low field signal broadens and eventually collapses completely at about $-25°$. On further decreasing the temperature, two new bands appear, a broad asymmetric band at $\tau \sim 4$ of relative intensity 4 and a fairly broad band of relative intensity 1 at $\tau6.5$. Below $\sim -60°$ the broad band at $\tau \sim 4$ separates into two bands, each of relative intensity 2. This series of spectral changes is shown in Fig. 4-7.

The simple spectrum above room temperature has been interpreted as the result of the stereochemical nonrigidity of the molecule because of the rapid shifting of the iron–carbon σ-bond from one carbon atom of the ring to another. Because of the averaging of the proton environment, the $\sigma\text{-}C_5H_5$ group exhibits only one NMR signal at room temperature. At low temperature the spectrum is that of an HA_2B_2 system [7-13]. The lowest field signal has been assigned to the A protons on the basis of its less well-resolved multiplet structure compared to the band at a slightly higher field. This is expected for the A protons since they experience spin–spin coupling with both the H proton and the set of B protons.

[4-13]

Support for a rearrangement mechanism involving a series of 1,2 rather than 1,3 shifts in such stereochemical nonrigid molecules has been found in

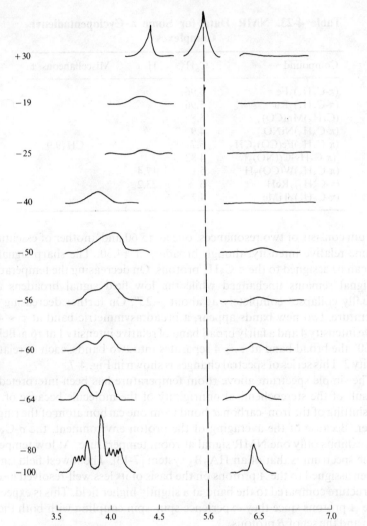

Fig. 4-7. Temperature dependence of the NMR spectrum of $[(\pi\text{-}C_5H_5)Fe(CO)_3(\sigma\text{-}C_5H_5)]$ in CS_2 solution.

the variable temperature NMR study of the corresponding σ-bonded indenyl compound, $(\pi\text{-}C_5H_5)Fe(CO)_2(1\text{-indenyl})$. Hückel LCAO-MO calculations have shown that a series of 1,2 shifts is very energetically unfavorable for the Fe-σ-indenyl structure. Such a mechanism would have to follow the course shown in equation (4-1). In II the type of π-electron distribution is ~9 kcal/ mole less stable than in I. On the other hand, if the pathway involves 1,3

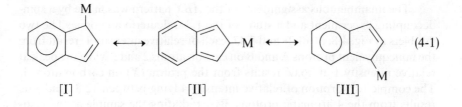

$$(4-1)$$

[I]   [II]   [III]

shifts, with a π-allyl-like transition state as shown in equation (4-2), the indenyl system should display rapid rearrangement since formation of the intermediate IV is not significantly hindered by the presence of the fused benzene ring. If a 1,3 shift were operative here, the five-membered ring of the

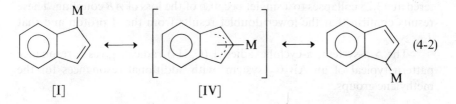

$$(4-2)$$

[I]   [IV]

indenyl system would be expected to give an A_2X spectral pattern, the protons on C atoms 1 and 3 having a time average equivalence. However, if the 1,2 shift mechanism provides the only significant pathway for the rearrangement, the molecule should be stereochemically rigid and exhibit an ABX pattern for the side ring as well as a complex absorption because of the aromatic protons. The latter is experimentally observed, with no detectable broadening up to 70°. The NMR of $(\pi\text{-}C_5H_5)Fe(CO)(1\text{-indenyl})$ is shown in Fig. 4-8.

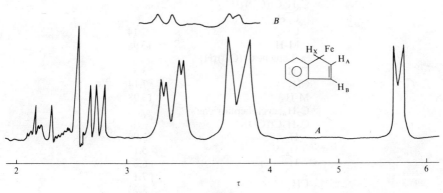

Fig. 4-8.

The unambiguous assignments of the ABX pattern was made by a spin-decoupling experiment and a study of the 1,3-dideuterio analogue. The two doublets in Fig. 4-8, at $\tau 3.28$ and 3.47, each of relative intensity 1, result from the nonequivalent protons A and B on carbon atoms 2 and 3 while the peak at relative intensity 1 at $\tau 6.06$ results from the proton (X) on carbon atom 1. The complex absorption of relative intensity 4 lying between $\tau 2.35$ and 3.05 results from the 4 aromatic protons. By irradiating the sample at the resonance frequency of the X proton (spin decoupling) the secondary splitting $(J = 2.5$ cps) of the lower τ doublet is removed, (Fig. 4-8B). The results of this double resonance experiment establishes that the X proton is coupled to either the A or B proton by 2.5 cps (but not to both). On replacing H_x and H_B by deuterium, the upper doublet ($\tau 3.47$) disappears while the one centered at $\tau 3.28$ collapses to a singlet because of the loss of AB coupling. These results confirm that the lower doublet results from the A proton and that $J_{AX} \gg J_{BX}$.

The 5 π-system, π-cyclohexadienyl (C_6H_7) radical, gives a resonance pattern typical of an AB_2C_2 system, with additional resonances for the methylene groups.

Table 4-24. NMR Data for Arene π-Complexes and for Some Free Arenes

Compound	τ
$C_6H_6FeC_6H_8$	5.13
$C_6H_6RuC_6H_8$	5.0
$C_6H_6Cr(CO)_3H^+$	
$\quad C_6H_6$	
$\quad CH_2$	7.14
$\quad$ M-H	13.98
$[\pi$-Toluene $Cr(CO)_3H]^+$	
$\quad C_6H_5$	3.35
$\quad CH_2$	7.14
$\quad$ M-H	13.98
C_7H_8, cycloheptatriene	
$C_7H_8Cr(CO)_3$	
$\quad H_a$	3.99
$\quad H_b$	5.17
$\quad H_c$	6.6
$\quad CH_2$	$\begin{cases} 7.1 \\ 8.23 \end{cases}$

Arene (6 π-System)

Arene π-complexes also give a single resonance peak in the same general region as observed for π-cyclopentadienyl systems (Table 4-24). As is noted in the table, the cycloheptatriene system is a 6 π-system with three types of protons plus methylene proton resonance.

4-3. MASS SPECTRA

The application of mass spectroscopy for the characterization of metal π-complexes is becoming increasingly popular. This section deals with the mass spectral data of several π-cyclopentadienyl, π-arene, and olefin metal π-complexes. Extensive studies have been made on π-cyclopentadienyl complexes, but little work has been done on the correlation of the mass spectral cracking patterns and energies of π-arene and π-allyl metal complexes.

For biscyclopentadienyl complexes of Va, Cr, Fe, Co, Ni, and Ru, the base peak in the mass spectra was the parent molecule ion. The C_5H_5 is not extensively fragmented, and there is a tendency to loose C_5H_5 units. In contrast, the mass spectra of $Mn(C_5H_5)_2$ gives more intense lower mass fragments. This has been interpreted as reflecting a greater ionic bonding in $Mn(C_5H_5)_2$ than in the corresponding complexes of the above metals.

Table 4-25 gives mass spectra data for ferrocene. Several features of the electron bombardment of ferrocene are as follows:

1. It appears that the parent molecule ion was formed by the removal of an electron localized on the iron atom.
2. As noted earlier, the cyclopentadienyl groups are removed intact and rather easily because of the large intensities of $Fe(C_5H_5)_2^+$, $FeC_5H_5^+$, and Fe^+.
3. $FeC_3H_3^+$ is present, probably as the result of the elimination of acetylene from $FeC_5H_5^+$.
4. Evidently the iron atom can be removed from the complex, resulting in the formation of $C_{10}H_9^+$ and $C_{10}H_8^+$ ions.

Mass spectrographic data for dibenzenechromium is given in Table 4-26. It is noteworthy that a large number of metastable species was observed that apparently formed by thermal dissociation of $Cr(C_6H_6)_2$.

The mass spectrum of cyclobutadieneiron tricarbonyl corresponds to the emission of the parent molecular ion and of derived species having one, two, and three fewer carbonyl groups. Peaks corresponding to the loss of C_2H_2 fragments were not observed. This type of data would suggest a structure such as [4-14] over any structure similar to [4-15].

Table 4-25. Mass Spectra of Ferrocene

Ion	$C_3H_3^+$	Fe^+	FeH^+	$C_5H_5^+$	FeC_2H^+	$Fe(C_5H_5)_2^{2+}$	$Fe(C_5H_5)_2^{2+} + FeC_3H_2^+$	$FeC_3H_2^+$
Relative abundances	2.5	13.1	0.4	0.7	2.1	3.8	2.4	3.5

Ion	$FeC_5H_5^+$	$C_{10}H_8^+$	$C_{10}H_9^+$	$C_{10}H_{10}^+$	$Fe(C_5H_5)_2^+$
Relative abundances	25.0	1.5	1.5	0.4	100.0

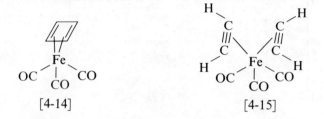

[4-14]   [4-15]

Table 4-26. Mass Spectra of Dibenzenechromium

m/q	70 eV relative abundance		m/q	70 eV relative abundance	
38	3.6	(5.5)	79	8.8	(6.4)
39	13.7	(13.0)	91	6.4	
41	6.3		42	1.7	
43	9.2		104	7.6	
44	5.0		104.5	0.9	} $Cr(C_6H_6)_2$
50	22.5	(15.5)	117	4.4	
51	18.2	(18.3)	118	1.4	
52	87.8	(18.9) } Cr	128	4.5	
53	8.4	(0.9)	129	4.9	
54	2.1		130	60.5	} $Cr(C_6H_6)$
55	8.3		131	12.2	
57	7.0		132	2.6	
63	4.8		160	4.0	
65	1.8		206	3.6	
69	5.4		207	2.1	
74	6.4	(4.6)	208	63.4	} $Cr(C_6H_6)_2$
76	5.5	(5.9)	209	16.3	
77	21.8	(14.7)	210	39	
78	100.0	(100.0)			

4-4. MÖSSBAUER SPECTROSCOPY

Mössbauer's discovery of gamma-ray resonance fluorescence is rapidly developing into a useful structural tool. The technique has the advantages of being highly precise and relatively easy to carry out. The position of the resonance and its splitting pattern can give significant information about electron densities, bonding, and site symmetries as well as the magnitude and direction of electric field gradients.

Although the Mössbauer effect has been observed for 32 elements, almost half of the work done in this area has been with iron. Most of the

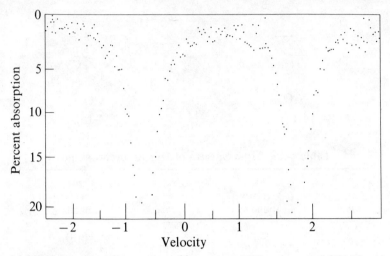

Fig. 4-9. Mössbauer absorption spectrum of ferrocene at 4.2°K and a zero-G field. The velocity relative to iron in millimeters per second.

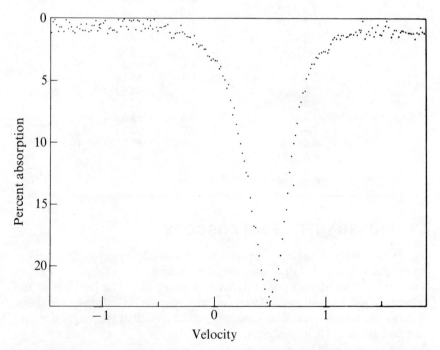

Fig. 4-10. Mössbauer absorption spectrum of ferrocenium dichlorodicyanoquinone at 77°K. Velocity relative to iron in millimeters per second.

investigations involve long-known inorganic coordination complexes. However, recently the Mössbauer spectra of ferrocene and ferrocinium ion have been reported (Figs. 4-9 and 4-10).

Ferrocene exhibits a quadruple splitting consistent with its nuclear spin moment of $\frac{1}{2}$. While the ferrocinium ion gives the same chemical shift as ferrocene (0.65mm/sec), it does not exhibit quadrupole splitting. This singlet structure has been explained by a molecular orbital treatment in which the removal of one electron from the highest bound state collapses the splitting by a fortuitous compensation of the large electric field gradient of the σ- and π-orbitals on the one hand and of the σ-orbitals on the other.

In spite of the strong screening of the s electrons expected by the ten d electrons, the observed chemical shift for ferrocene is relatively small. This suggests that the four s electrons participate to about 30 % in the bonding or there is π-bonding from the iron to the ring involving the d_{yz} and d_{xz} orbitals.

Only small changes in the spectral parameters are observed on ring substitution, although the field gradient is often slightly reduced from 2.70 mm/sec to 2.6 mm/sec. The only case of a shift to higher field is when a CH within the ring is substituted by nitrogen to give the heterocyclic azaferrocene, $C_5H_5FeC_4H_4N$ ($\Delta = 2.74$ mm/sec).

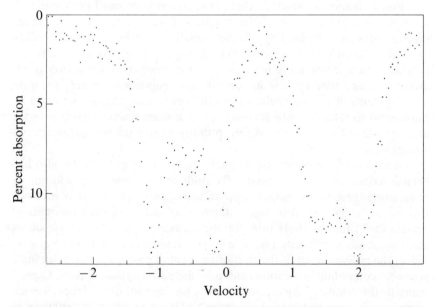

Fig. 4-11. Mössbauer absorption spectrum of ferrocene at 4.2°K and 40,000 G. Velocity relative to iron in millimeters per second.

A recent study has established that the electric field gradient in ferrocene is negative by measuring its Mössbauer spectrum under a magnetic perturbation (Fig. 4-11). These results support the prediction of MO calculations, CFT having predicted a positive electric field gradient.

4-5. MAGNETOCHEMISTRY

Magnetic measurements find intensive applications in the study of transitional metal π-complexes. Such data can give information that is useful for the elucidation of valency, bond type, and stereochemistry of such compounds.

Magnetic effects arise from the motion of electrons regarded as charged particles (diamagnetism) and from spin and orbital angular momentum of electrons (paramagnetism). Experimentally the Gouy method is most commonly used to measure magnetic susceptibility. In this method the weight change of a sample under the influence of an applied homogeneous magnetic field is measured by the use of a calibrated spring or an accurate electrobalance. At least a few tenths of a gram of the substance under study is required for this technique.

For diamagnetic materials this parameter is very small in magnitude, negative, and independent of the magnetic field and temperature. Paramagnetism is characterized by relatively small, positive susceptibility values that are independent of magnetic field strength but display an inverse dependence on a temperature term. Since diamagnetism is a property of all atoms whether they possess an overall paramagnetism or not, accurate measurements of paramagnetic susceptibilities require that a correction be introduced to take this into account. Thus, a diamagnetic correction term must be added to the measured susceptibility to give the real paramagnetic susceptibility.

A method for determining magnetic susceptibilities by NMR also has been described and is widely used in the study of π-complexes. In addition to eliminating the rather expensive apparatus required for the Gouy method, this technique has the advantage of the need of only very small amounts of sample (5–20 mg), a short time for the measurement, and the use of an instrument now commonly available in most laboratories. The technique involves the measurement of the shift of the proton resonance line of an inert reference compound in solution caused by the paramagnetic species. Experimentally this is most easily accomplished by having a solution of the reference compound in a concentric tube inside the NMR tube containing a solution of the reference compound and paramagnetic species. The solvent itself also can be used as the reference compound. The mass susceptibility X of the dissolved sample is then given by equation (4-3), where Δf is the frequency,

$$X = \frac{3\Delta f}{2\pi fm} + X_o + \frac{X_o(d_o - d_s)}{m} \qquad (4\text{-}3)$$

separation in cps between the two lines, f is the frequency at which the proton resonances are being studied in cps (60×10^6 cps for the most common commercial instrument), m is the mass of sample contained in 1 ml of solution, X_o is the mass susceptibility of the solvent, d_o is the density of the solvent, and d_s is the density of the solution. For highly paramagnetic substances the last term is generally neglected without serious error.

Metal π-complexes generally assume a low spin configuration. This magnetic property is the result of the strong LF (enhanced covalence in MOT) characteristic of ligands in such compounds and reflected in the tendency to form a closed shell (noble gas configuration) around the metal. The stabilization of a lower oxidation state in π-complexes may be explained qualitatively by the back donation of electrons from a zero-valent metal into the vacant antibonding orbitals on the π-complexed olefin. The situation is the same in metal phosphine compounds where the vacant d-orbitals of the phosphorus play the same role in stabilizing a lower oxidation state.

The magnetic moments (Table 4-27) of arene type complexes reveal that the formation of complexes generally results in diamagnetism and the closest possible approach to the next "inert gas configuration."

As stated above, the formation of metal π-complexes usually results in the stabilization of the lower valence states of the metal. For example,

Table 4-27. Magnetic Moments of Diarene Metal π-Complexes

Metal (oxidation state)	Electron configuration	Compound	Magnetic moment (BM)
Cr(1)	d^5	$Cr(C_6H_6)_2^{+1}$	1.73*
V(0)		$V(C_6H_6)_2$	1.68
Cr(0)	d^6	$Cr(C_6H_6)_2$	Diamagnetic
Mo(0)		$Mo(C_6H_6)_2$	Diamagnetic
W(0)		$W(C_6H_6)_2$	Diamagnetic
Fe(2)		$[\text{sym-}(CH_3)_3C_6H_3\,_2Fe]^{2+}$	Diamagnetic
Ru(2)		$[\text{sym-}(CH_3)_3C_6H_3\,_2Ru]^{2+}$	Diamagnetic
Os(2)		$[\text{sym-}(CH_3)_3C_6H_3\,_2As]^{2+}$	Diamagnetic
Rh(3)		$[\text{sym-}(CH_3)_3C_6H_3\,_2Rh]^{3+}$	Diamagnetic

* The calculated magnetic moment for a first transition row element having one unpaired electron is 1.73 BM.

chromium, which ordinarily exhibits principle oxidation states of $+6$, $+3$, and $+2$, is stabilized at oxidation states of $+1$ and 0. Thus dibenzechromium (oxidation state 0) is thermodynamically very stable in an inert gas atmosphere and melts at 284–285°. In the presence of oxygen and water it is readily oxidized to dibenzechromium(I) cation, which is very stable in air and can be readily isolated in the solid state with PF_6^- or other bulky anions (Table 4-28).

The magnetic moments of metallocenes and di-π-cyclopentadienyl metal complexes are very similar to diarene metal complexes. The magnetic data of some metallocenes are shown in Table 4-29.

Table 4-28. Oxidation State of Metals in Arene π-Complexes

Metal	Major oxidation states* in inorganic compounds	Oxidation states in arene π-complexes
V	5, 4, 3, 2	0, 1
Cr	6, 4, 3, 2	0, 1
Mo	6, 5, 4, 3, 2	0, 1
W	6, 5, 4, 3, 2	0, 1
Mn	7, 6, 4, 3, 2	1
Tc	7	1
Re	7, 6, 4, 2, −1	0, 1
Fe	3, 2	0, 1, 2
Ru	8, 6, 4, 3, 2	0, 2
Co	3, 2	1, 2, 3
Rh	4, 3, 2	0, 1, 2, 3

* Underlined number denotes the common oxidation state.

Table 4-29. Magnetic Data of Some Metallocenes

Complex	Eff, BM
π-Cp$_2$Ti	Diamagnetic
π-Cp$_2$V	3.82
π-Cp$_2$Cr	2.84
π-Cp$_2$Fe	Diamagnetic
π-Cp$_2$Co	Diamagnetic
π-Cp$_2$Ni	2.88
[π-CpCr(CO)$_3$]$_2$	Diamagnetic
[π-CpFe(CO)$_2$]$_2$	Diamagnetic

The electronic configuration of metal π-complexes that contain carbonyl groups also tend to form a closed shell, a formal "noble gas configuration."

All the tricyclopentadienyl derivatives of the lanthanides, $Ln(C_5H_5)_3$, exhibit the same "free ion" value observed in their simple inorganic salts. The uranium derivative, $(\pi\text{-}C_5H_5^-)_3UCl$, has a magnetic moment of 3.16 BM (Bohr magnetons) in the solid state, indicating two unpaired $5f$ electrons.

4-6. EXERCISES

4-1. What is the characteristic NMR pattern of π-allyl groups?

4-2. Does an NMR signal appear with paramagnetic π-complexes?

4-3. Distinguish the following allyl groups with respect to their NMR spectra: σ-allyl, π-allyl, and dynamic allyl.

4-4. What change in infrared spectrum is expected when a terminal olefin is π-complexed with $PtCl_2$?

4-5. How much shift in $C\equiv C$ stretching frequency is expected when an acetylene, for example, $R-C\equiv CH$, is π-complexed with $Co(CO)_3$ groups or with a $Pt(PPh_3)_2$ group?

4-6. List the characteristic absorption bands of a covalently bonded π-cyclopentadienyl group.

4-7. What is the difference in the infrared spectrum between ionic cyclopentadienyl, cyclopentadienide, and π-cyclopentadienyl groups?

4-8. What kind of shift in NMR spectra is expected for a conjugated diene upon metal π-complex formation?

4-9. Does the sharp, single NMR peak at room temperature unambiguously indicate a planar, π-bonded cyclooctatetraene ring in $C_8H_8Fe(CO)_3$? Discuss.

4-10. Which protons are most strongly shielded in butadiene iron tricarbonyl, $C_4H_6\,Fe(CO)_3$?

4-11. In what range does the NMR signal of π-cyclopentadienyl protons appear?

4-12. Give examples of valence tautomerism in various π-complexes.

4-13. What kind of change in the $C-H$ stretching frequency ($vC-H$ of the $\diagdown C\diagup CH_2$ group is expected when cyclopentadiene is coordinated with a $Co(\pi\text{-}C_5H_5)$ or a $Rh(\pi\text{-}C_5H_5)$ group?

4-14. What is the reason for the change of $C=C$ stretching vibration ($vC=C$) on π-complex formation?

4-15. On what elements or its compound can the Mössbauer effect be observed most easily?

4-16. List the characteristic absorption bands of π-benzene metal complexes.

4-17. What kind of information can be obtained from the electronic (ultraviolet and visible) spectra of π-complexes?

4-18. Summarize the changes observed in the following properties upon π-complex formation:

a. Conjugation   c. Conformation
b. Acidities and basicities   d. Bond lengths

4-19. Microanalytical results indicate that the molecule formula of a compound is $PdClC_3H_5$. Molecular weight determinations give 362.5. The compound in the double-bond region of the IR exhibits a band at 1458 cm^{-1}. Treatment of this compound with four equivalents of triphenylphosphine gives a compound whose molecular weight was found to be 751.5 and which gave a band at 1630 cm^{-1} in the IR. Give the structures of both compounds.

4-7. BIBLIOGRAPHY

"Advances in Organometallic Chemistry," Vols. 1–4, Academic Press, New York (1964, 1964, 1965, 1966).

"Annual Surveys of Organometallic Chemistry," Vols. 1 and 2, Elsevier Publishing, Amsterdam (1965 and 1966).

F. A. Cotton, "The Infrared Spectra of Transition Metal," in "Modern Coordination Chemistry Complexes" (J. Lewis and R. G. Wilkins, eds.), Interscience Publishers, New York (1960).

G. R. Dobson, I. W. Stolz, and R. K. Sheline, "Substitution Products of the Group VIB metal Carbonyls," in "Advances in Inorganic Chemistry and Radiochemistry" (H. J. Emeleus and A. G. Sharpe, eds.), Academic Press, New York (1959).

R. S. Drago, "Physical Methods in Inorganic Chemistry," Reinhold Publishing, New York (1965).

B. N. Figgis and J. Lewis, "The Magratochemistry of Complex Compounds," in "Modern Coordination Chemistry" (J. Lewis and R. G. Wilkins, eds.), Interscience Publishers, New York (1960).

E. Fluck, "The Mössbauer Effect and its Application in Chemistry," in "Advances in Inorganic Chemistry" (H. J. Emeleus and A. G. Sharpe, eds.), Academic Press, New York (1964).

D. K. Higgins and H. D. Kaesg, in "Progress in Solid State Chemistry" (H. Reiss, ed.), The Macmillan Company, New York (1964).

M. Kubo and D. Nakamura, "Nuclear Quadrupole Resonance and Its Application in Inorganic Chemistry," in "Advances in Inorganic Chemistry and Radiochemistry (H. J. Emeleus and A. G. Sharpe, eds.), Vol. 8, p. 257, Academic Press, New York (1966).

K. Nakamoto, "Infrared Spectroscopy," R. H. Herber, "Mössbauer Spectroscopy," and R. Keiser, "Mass Spectroscopy," in "Characterization of Organometallic Compounds" (M. Tsutsui, ed.), Interscience Publishers, New York (in press).

G. Wilkinson and F. A. Cotton, "Cyclopentadienyl and Arene Metal Compounds," in "Progress in Inorganic Chemistry" (F. A. Cotton, ed.), Vol. I, p. 1, Interscience Publishers, New York (1959).

CHAPTER 5

Structure and Structure Determination

The area of the structural elucidation of compounds has made giant strides since a few decades ago when the chemist was generally restricted to the use of chemical reactions, such as ozonolysis, hydrogenation, and melting points, in establishing the nature of newly synthesized materials. Although some may lament the current trend away from pure chemistry and toward physical methods, one cannot help but be impressed by the array of tools available to the modern chemist and the vast amount of detailed structural information resulting from their knowledgeable application.

The area of metal π-complexes is especially susceptible to modern structural studies since both the physical probes originally applied to organic and inorganic chemistry can be utilized here.

The standard techniques of infrared spectroscopy and nuclear magnetic resonance spectroscopy have been most widely applied in the structural elucidation of metal π-complexes. Electronic spectroscopy, electron spin resonance spectroscopy, conductivity, and dipole moment measurements also have been frequently used. Mass spectroscopy and Mössbauer spectroscopy are being increasingly used to obtain detailed information on the bonding in metal π-complexes.

Of course the most valuable structural tool is the single crystal x-ray crystallography study. Because of the experimental difficulties and time-consuming nature of such studies, however, every new compound cannot be the object of such a structural determination. The method, rather, is to establish the structure of a model compound by x-ray analysis and then to assign analogous structures to other compounds on the basis of the comparison of more easily obtainable spectral data, such as infrared and nuclear magnetic resonance. This discussion concentrates on structures themselves by introducing isomerism encountered in metal π-complexes.

5-1. ISOMERISM IN METAL π-COMPLEXES

Molecules or ions having the same chemical composition but different structures are called isomers. Metal complexes exhibit several different types of isomerism: The three most important are linkage (see σ–π rearrangements in Chapters 4 and 6), geometric, and optical. Since isomerism in metal π-complexes is governed in principle by the nature of both the metal atom and the organic π-ligands, some complexity of stereochemistry is expected. However, investigation in this area of π-complex chemistry is limited. In the past progress has been directed mainly toward synthesis and characterization.

Geometrical

In metal complexes the ligands may occupy different positions around the central atom. Since the ligands in question are usually either next to one another (*cis*) or opposite each other (*trans*), this type of isomerism is often referred to as *cis–trans* isomerism. *Cis–trans* isomerism is very common for square planar and octahedral complexes. Consider the square planar complexes shown in structures [5-1]–[5-4]. *Cis–trans* isomerism arises from the relative position of the ethylene ligands. Therefore, [5-1]* and [5-3] are *cis* forms and [5-2] and [5-4] are *trans* forms, respectively. Norbornadiene-palladium dichloride [5-5] and rhodium (π-cycloocta-1,5-diene) chloride [5-6] can exist only in the *cis* structure because of the chelate nature of the diene.

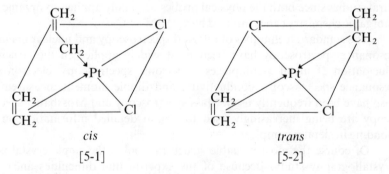

	cis	*trans*
	[5-1]	[5-2]

* Double numbers in brackets refer to structural formulas.

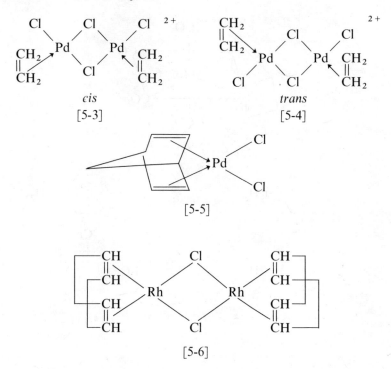

[5-3] *cis*

[5-4] *trans*

[5-5]

[5-6]

Cis–trans isomerism in octahedral complexes is shown in structures [5-7] and [5-8]. Norbornadiene molybdenum tetacarbonyl can assume only the *cis* form, as seen in [5-9]. Arene π-complexes also may exhibit *cis–trans* isomerism. X-ray diffraction data have shown that of the two structures [5-10] and [5-11], which could exist for ditoluenechromium(0), the thermodynamically more stable structure [5-11] exists under ordinary conditions. Dibiphenylchromium(0) [5-12] and [5-13] may also exist in at least two forms.

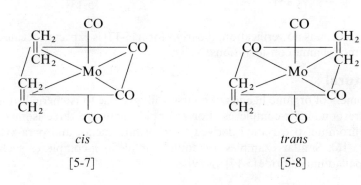

[5-7] *cis*

[5-8] *trans*

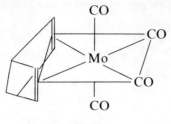

[5-9]

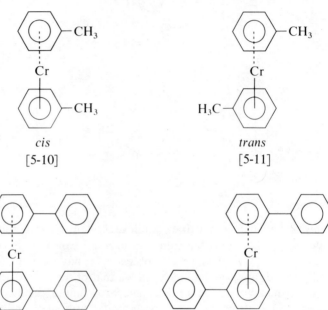

cis	trans
[5-10]	[5-11]

cis	trans
[5-12]	[5-13]

Although there is no verification, the *trans* form [5-13] is expected because of the thermodynamic considerations.

Structural

Isomers of organic ligands of course will give rise to isomeric forms of the corresponding π-complexes. For example, there are three isomers of xylenechromium tricarbonyl derived from ortho-, meta-, and para-xylene [5-14]–[5-16]. Similar examples are found for olefin π-complexes such as butenepalladium chloride [5-17]–[5-19].

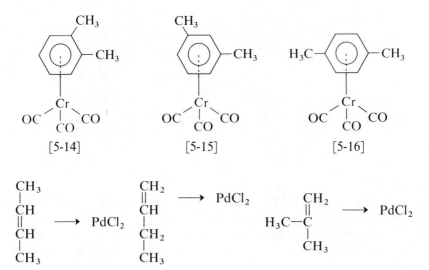

[5-14]   [5-15]   [5-16]

[5-17]   [5-18]   [5-19]

Conformational

Theoretically, metallocenes and arene π-complexes can exhibit conformational isomerism. The cyclopentadienyl rings in ferrocene should exist in two extreme conformations, either eclipsed [5-20] (prismatic ferrocene) or staggered [5-21] (antiprismatic ferrocene). According to x-ray diffraction

[5-20]   [5-21]

data, the rings are staggered [5-21] and ferrocene is antiprismatic. However, it is interesting to note that both ruthenocene and osmocene are prismatic and that the cyclopentadienyl rings therefore are eclipsed. The difference in conformation is considered to be because of the larger sizes of ruthenium and osmium compared to iron, resulting in less sterric strain in the former cases. Similarly, the benzene rings in dibenzenechromium(0) could be either staggered or eclipsed [5-22] or [5-23]. The eclipsed structure [5-23] has been confirmed by x-ray diffraction.

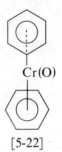

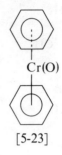

[5-22]   [5-23]

Optical

As in organic chemistry, optical isomerism in metal π-complexes arises from a lack of symmetry, referred to as asymmetry.

One type of optical isomerism is exhibited simply by the incorporation of an asymmetric organic ligand in a metal π-complex, as shown in structure [5-24].

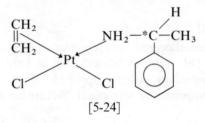

[5-24]

Complexes [5-25]–[5-27] are good examples of asymmetric π-complexes that possess molecular asymmetry. They have been successfully resolved into enantiomers (mirror images).

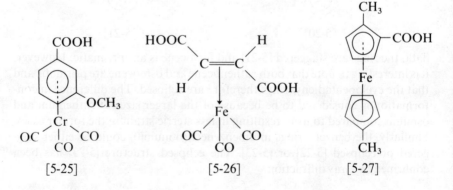

[5-25]   [5-26]   [5-27]

Although never prepared, structure [5-28] represents an interesting type of metal π-complex. The central metal M is asymmetric and analogous to the asymmetric carbon in structure [5-24]. Therefore, optical activity is predicted. Investigation of this field has been stimulated by the development of the stereospecific polymerization of olefins by the Zieggler–Natta catalysts.

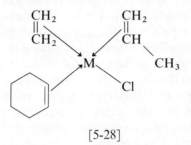

[5-28]

Since these catalysts are transition metal complexes, the mechanism of stereospecific polymerization most likely involves transitory π-complex formation between the olefin and the catalyst.

5-2. DETERMINATION OF STRUCTURE

Elemental analysis and the determination of molecular weight are as essential in proposing the structure of a new π-complex as they are in the elucidation of a simple new organic or inorganic compound. The unequivocal assignment of a π-bond between a ligand and a metal atom is the most difficult feature of structural determination since, as discussed in Chapter 4, spectral data often can be ambiguous. The existence of such a bond has been deduced indirectly by the interpretation of chemical and physical data. Complete characterization usually requires examination by a number of the following methods: chemical reactivity, chemical degradation, preparation of derivatives, visible and ultraviolet (UV) spectra, infrared (IR) spectrum, nuclear magnetic resonance (NMR), optical rotary dispersion, magnetic susceptibility measurement and electron paramagnetic resonance (EPR), Mössbauer spectrum, and dipole moment. Examination of the UV absorption spectrum can provide data on both the electronic configuration of the π-ligand and the metal atom in a π-complex. A study of the dipole moment, Mössbauer spectrum, UV absorption spectrum, magnetic susceptibility, and EPR provide an indication of the formal valence state of the metal in a complex as well as detailed information covering charge density on the ligands. The dipole moment and IR spectrum can be used to deduce the stereochemistry of

the compound. Chemical methods also play an important role in the characterization of the structure.* The single crystal x-ray study, however, is the
ultimate structural tool. With rare exception x-ray crystallography reveals the
overall geometry of a compound as well as quite accurate measurements of
individual bond angles and bond lengths. In cases where ambiguity arises
from x-ray data, generally as the result of the presence of two atoms of very
similar size such as C and N, neutron diffraction studies generally can be
successfully applied to supplement this data and give absolute structures. In
the case of dibenzenechromium the results of x-ray diffraction analysis per-
formed by different investigators have created some controversy concerning
the C—C bond length in the benzene rings, as described later in this chapter.
This may be due to experimental errors or procedural limitations. Neutron
diffraction analysis may provide a more refined determination of the C—C
bond length. In this section three instructive examples of structure elucidation
of metal π-complexes are presented and the structures of analogous metal
π-complexes determined by x-ray diffraction analysis are illustrated.

Zeise's Salt

In 1829 a crystalline substance was isolated by refluxing platinum(IV)
chloride in ethanol and then adding potassium chloride. This compound,
Zeise's salt, was analyzed as $KCl—PtCl_2C_2H_4$. Upon thermal decomposition, some ethylene was liberated. Hydrolysis gave acetaldehyde. Because of
the liberation of ethylene and the oxidative hydrolysis to acetaldehyde, it was
suspected that ethylene was somehow incorporated into this apparent platinum chloride salt. Many tentative structures were proposed, for instance
[5-29] and [5-30].

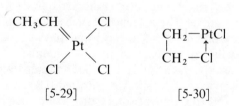

[5-29]   [5-30]

Liebig, the most distinguished chemist of that time, severely criticized
Zeise's work, largely on the basis that the bonding, the analysis, and the
chemical behavior were not reconcilable.

* Several chemical ractions are commonly applied as diagnostic criteria of the bonding
mode of the cyclopentadienyl group. The formation of ferrocene by the reaction of ferrous
chloride with a new complex is taken as evidence of the presence of an ionic cyclopentadienyl
group. Similarly, reaction with maleic anhydride to give the Diels–Alder product indicates
either an ionic or σ-bonding mode for the cyclopentadienyl group rather than π-bonding.

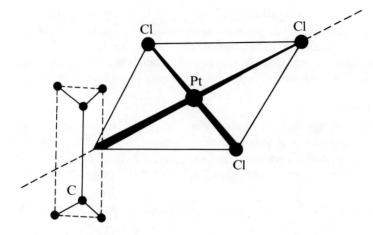

Fig. 5-1. Square planar representation of the $[(C_2H_4)PtCl_3]^-$ anion.

When ferrocene was discovered, the concept of a metal π-complex bond was introduced. Thus an explanation of the ethylene to platinum π-complex bond may be proposed. Generally, Pt^+ with a d^8 electron configuration takes a square planar configuration dsp^2. Thus the structure of Zeise's salt was proposed largely by the IR spectral studies, as shown in Fig. 5-1.

Subsequent x-ray diffraction analysis proved the square planar configuration (Fig. 5-1) to be correct. Furthermore, it was revealed that the three chloride atoms are in the same plane as the platinum, while the ethylene molecule is perpendicular to the plane, as shown. A slight distortion of the basic square planar structure is noted since the platinum to chlorine bond length *trans* to ethylene is longer than those that are *cis*. A similar example of a square planar metal π-complex is seen in π-allylpalladium chloride $[(\pi\text{-}C_3H_5)PdCl]_2$ (Fig. 5-2).

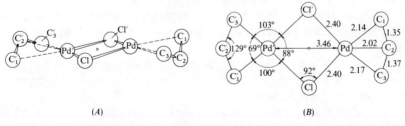

(A)   (B)

Fig. 5-2. Di-π-allylpalladium chloride. *A*: The molecular configuration of $[(C_3H_5)PdCl]_2$. *B*: Intra molecular distances and angles.

The evidence for π-bonding is based on the fact that the plane of the allyl group is approximately perpendicular to the plane of the two palladium and two chlorine atoms. Molecular weight determinations have confirmed its formation as a dimer while IR spectral studies established the presence of bridging chloro groups as opposed to terminal groups.

Ferrocene

An orange-colored organo-iron compound A was isolated from the reaction mixture of sodium cyclopentadienide (3 moles) and iron (III) chloride (1 mole) in ether. A was analyzed as $C_{10}H_{10}Fe$. Since A was very stable towards acids and bases, structures [5-31] and [5-32] originally proposed were not acceptable.

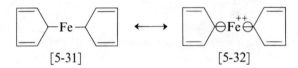

[5-31]   [5-32]

The iron-C bonds in [5-33] and [5-34] should be readily cleaved by acids or bases.

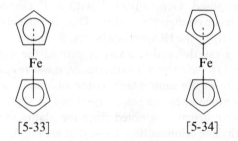

[5-33]   [5-34]

A is readily oxidized to a blue cation $[Fe(C_{10}H_{10})]^+$ by acids and exhibits the following properties: dipole moment, zero; magnetic moment, diamagnetic; and solubility, very soluble in hydrocarbons.

Without benefit of NMR data, a pentagonal antiprismatic structure [5-33] was proposed. However, the evidence given above also was true for the prismatic structure [5-34].

Later the NMR of ferrocene was studied. A singlet signal at 5.69τ suggested that the iron atom is located in the center of the two cyclopentadiene rings; thus structure [5-33] or [5-34] is reasonable.

Subsequent x-ray diffraction analysis (Fig. 5-3) supported the structure [5-33] in which the π-cyclopentadienyl rings are in a staggered position. However, it should be noted that in solution at room temperature the rings are

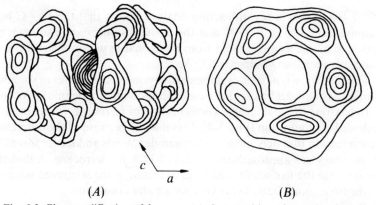

Fig. 5-3. Electron diffration of ferrocene. A: Superposition of a section from the electron density function for ferrocene giving a composite picture of the staggered rings sandwiching the iron atom. B: A section of the electron density function in the plane of one of the cyclopentadiene rings. Reprinted with permission from J. D. Dunitz, L. E. Orgel, and A. Rich, *Acta Cryst.* **9**:374 (1956).

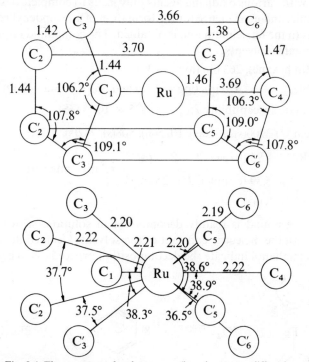

Fig. 5-4. The structure of rutheneocene (based on x-ray diffraction data). Reprinted with permission from G. P. Gardgrave and D. H. Templeton, *Acta Cryst.* **12**:28 (1959).

rotating freely. The x-ray diffraction analysis shows that the Fe—C bonds are equivalent 2.045 ± 0.01 Å, and the C—C distances are also equivalent (1.403 ± 0.02 Å), as anticipated from a trigonal bypyramid structure. The distance between the two rings is 3.32 Å.

The sandwich-type di-π-cyclopentadienyl metal complexes of vanadium, chromium, cobalt, and nickel are isomorphous with monoclinic ferrocene.

In contrast to the staggered configuration of ferrocene, ruthenocene has an eclipsed configuration (Fig. 5-4). This may be because of the difference in the atomic radii that may influence the van der Waals and lattice forces. The ring distances are approximately 3.68 vs 3.32 Å for ferrocene. Although in the solid state the five-membered rings are fixed in the staggered configuration, the rings rotate freely in the vapor state and in solution.

Bis-Indenyliron(II) (Dibenzferrocene)

This deep purple compound, prepared from the reaction of iron(II) chloride and sodium indenide in THF, is moderately air stable. Attempts to oxidize it with various oxidizing agents only leads to complete destruction of the molecule, and no evidence for the formation of a bis-indenyliron(III) ion, analogous to the ferricinium ion, is obtained. The compound is quite soluble in polar organic solvents:

UV (in hexane), 263 mμ (log −4.30)

Visible,   420 mμ (log −2.78) and 563 mμ (log −2.49)

Magnetic susceptibility $\lambda_{mol}^{297°K} = 175 \times 10^{-6}$ GSU

IR (cm^{-1}) 2900(s), 2850(s), 1025(m), 808(s), 742(s), 727(s)

NMR,   5.95 (triplet, $J = 2.5$ cps) ⎫
                                      ⎬ relative intensities are 1:2:4
         5.39 (doublet, $J = 25$ cps) ⎭

         3.009

The compound is readily decomposed in solution, with the rate increasing with the dielectric constant of the solvent. Detailed decomposition studies in methanol indicate the decomposition proceeds as in equation 5-1.

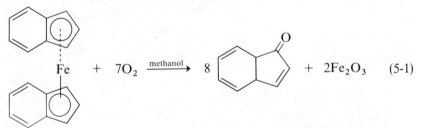

Hydrogenation of bis-indenyliron goes smoothly in ethanol with a PtO_2 catalyst to give bis- (tetrahydroindenyl) iron. The ease of hydrogenation, together with the volume of hydrogen absorbed (4 moles per mole of complex), demonstrates the olefinic character of the two double bonds in each six-membered ring.

On the basis of the above data, bis-indenyliron was formulated as having a *trans* sandwich-type molecular configuration [5-35].

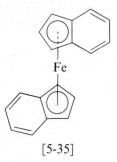

[5-35]

Subsequently, an x-ray crystallographic study showed that the molecular configuration was gauche with a perpendicular distance between the planes of the rings of 3.43 Å. This gauche configuration is possibly stabilized with respect to the *trans* by weak van der Waals forces between atoms of the two six-membered rings.

The ruthenium compound also has been prepared and displays similar properties, although it has not yet been the subject of an x-ray study. Bis-indenylcobalt, the first member of the series to be prepared, has a structure similar to the iron analogue on the basis of a comparison of x-ray powder photographs. This latter compound can be readily oxidized to give the quite stable Co(III) compound which, like the cobalticinium ion, may be precipitated as the triiodide, tribromide picrate, or perchlorate. The perchlorate exhibits a well-defined polarographic reduction wave at -0.6 V vs SCE. The corresponding half-wave potential for cobalticinium perchlorate is -1.16 V. Thus the annulation of a benzene ring produces a very marked decrease in the reduction potential and, conversely, the oxidation of the new compound to the ion should become more difficult.

π-Cyclopentadienyl-π-Pyrrolyliron

A small number of compounds containing a π-bonded pyrrolyl group has been prepared. Iron(II) chloride reacts with a mixture of sodium cyclopentadienide and sodium pyrrolide in THF to give red, very volatile, crystalline $(C_5H_5)Fe(C_4H_4N)$ (azaferrocene) in very low yield. This compound has also been prepared from potassium pyrrolide and $[(C_5H_5)Fe(CO)_2I]$. Since the volatility and solubility properties of this compound are quite similar to ferrocene, a major side product, it is isolated from the reaction mixture by use of an alumina column, on which it is much more strongly retained than ferrocene. It can be eluted by acetone, the ferrocene being readily removed by benzene:

> IR, CH band at 3090 cm$^{-1}(w)$; other bands at 1405(m), 1385(m),
>   1345(w), 1265(w), 1184(m), 1109(s), 1060(w), 1003(s), 853(w),
>   816(s), and 770(w)
> UV, 330 mμ (178) and a strong end absorption at 210 mμ (15,400)
> Visible, 442 mμ (107)
> PMR, τ4.74, 5.55 and 5.85 (relative intensities, 2:2:5)

This compound has been described as given in [5-36]. The singlet resonances at $\tau = 4.74$ and 5.55 are assigned to the two pairs of equivalent protons

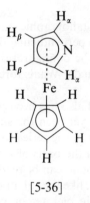

[5-36]

(bonded to α- and β-carbon atoms of the π-pyrrolyl ring). The compound $(C_4H_4N)Mn(CO)_3$, prepared by heating dimanganese dicarbonyl with pyrrole, shows a similar NMR spectrum but at lower field. This upfield shift of resonances of protons bonded to carbon atoms involved in π-bonding to a transition metal in compounds in which three carbonyl groups are replaced by a π-C_5H_5 ring seems to be a general phenomenon. For example, the π-C_5H_5 protons of $(C_5H_5)Mn(CO)_3$ have signals at $\tau = 5.35$, whereas the π-C_5H_5 protons of $(C_5H_5)_2$Fe occur at $\tau = 5.87$.

The absence of splitting between the two different types of protons in the spectrum of both these π-pyrrolyl complexes is not understood at present.

Dibenzenechromium(0)

From the reaction of benzene, anhydrous chromium(III) chloride, aluminum chloride, and aluminum powder, a bright yellow, crystalline, organochromium chloride, $C_{12}H_{12}CrCl$, B, was obtained. Reduction of the salt gave $C_{12}H_{12}Cr$, 2A, with the following physical properties: dipole moment, apparently zero; magnetic moment, $u = 0$ (compound B, $u = 1.79BM$); and IR, the normal skeletal vibrations were observed at 1430–1410 cm^{-1} and at 1140–1120 cm^{-1} rather than at 1600 and 1500 cm^{-1}, respectively.

Pyrolysis of one mole of A gives two moles of benzene. From the result of the pyrolysis, benzene is reasonably considered to exist in the compound. However, the IR spectra of the compounds revealed no benzene absorption; the compound is not a benzene occlusion complex. From this data, structure [5-37] was tentatively proposed. This structure can be reasonably explained by the valence bond theory, EAN, and molecular orbital theory to be an octahedral complex, as described in Chapter 3. Subsequently, structure [5-38] was assigned to the chloride salt B. From the limited data given above, it is not possible to assign the orientation of two benzene rings on the chromium atom.

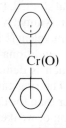

Cr(O)

Dibenzenechromium

[5-37]

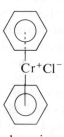

Cr$^+$Cl$^-$

Dibenzenechromium chloride

[5-38]

X-ray diffraction analysis of A revealed that two benzene rings are eclipsed. However, there have been intense arguments on the length of C—C bond in the benzene rings of [5-37].

Assuming a six-fold symmetry, the original chromium–carbon and carbon–carbon bond length were reported to be 2.19 ± 0.1 and 1.38 Å, respectively, by trial and error. A more refined analysis suggested alternating carbon–carbon bond lengths of 1.353 and 1.439 Å, respectively. This view of a threefold symmetry was in good agreement with a penetration complex theory in which electron pairs penetrate the chromium atomic orbitals in an octahedral complex, giving the benzene rings a cyclohexatriene nature. In a

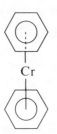

Fig. 5-5. The structure of dibenzenechromium(0).
C—C = 1.42 Å; Cr—C = 2.14 Å.

later study a third group reported that a sixfold symmetry is preserved and that the C—C bonds are essentially equivalent with bond lengths of 1.387 Å. In a refinement, the original investigators made two significant experimental changes. The study of the crystal structure was conducted at low temperature in order to suppress thermal motion, which puts a serious limitation on the accuracy with which the atomic positions can be determined at room temperature. In addition, since the difference in results may be because of orientational disorder in the crystal, dibenzenechromium(0) was prepared without the use of a mesitylene catalyst, which could have caused a partial substitution of benzene by mesitylene in the crystal studied. It was determined that there is no deviation from a D_{6h} molecular symmetry and that all the C—C bond lengths in dibenzenechromium are equivalent (Fig. 5-5).

Benzenechromium tricarbonyl provides an additional example of an octahedral π-complex (Fig. 5-6). The plane which is defined by the three oxygen atoms is parallel to the benzene plane. The ring carbon–chromium bond distances are 2.25 Å, which is slightly longer than the distance in dibenzenechromium, 2.19 Å. This may be an indication of weaker bonding of the benzene to the metal in the carbonyl complex. This is reasonable in view of the strong π-bonding ability of the carbonyl group, which would tend to decrease $Cr \rightarrow C_6H_6$ back bonding. The carbon and oxygen atoms of each carbonyl group are colinear with the chromium atom, and the angles OC—Cr—CO are equal and very nearly 90°, which shows an octahedral coordination of the CO groups. IR studies have confirmed such a C_{3v} symmetry for this compound.

Cyclo-octatetraene-Iron Tricarbonyl

Cyclo-octatetraene-iron tricarbonyl (A) was originally isolated from the reaction mixture of cyclo-octatetraene and iron pentacarbonyl. The properties of A are as follows: UV, $\lambda_{max} - 303$ mμ; increasing absorption at shorter wavelengths, and IR, 2058 and 1992 cm^{-1} (FeCO). Decomposition of A gave 1 mole of cyclo-octatetraene and 3 moles of CO.

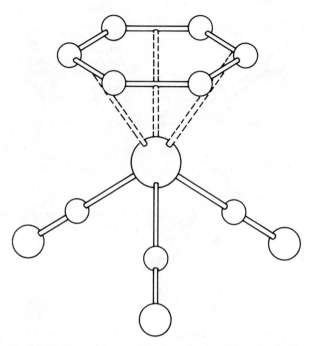

Fig. 5-6. Structure of benzenechromium tricarbonyl. Reprinted with permission from P. Corradine and G. Allegra, *Proc. Chem. Soc.* **81**:2271 (1959).

From the limited data above, the tentative structures [5-39] and [5-40] can be proposed. However, A does not absorb hydrogen even in the presence of active hydrogenation catalysts such as platinum or palladium, which would be expected to happen if two olefinic bonds were uncomplexed.

Furthermore, NMR reveals only one very sharp peak at 16.8 cps. This data tends to eliminate structures [5-39] and [5-40]. Thus structure [5-41] was

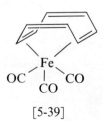

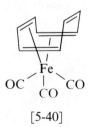

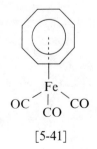

[5-39]   [5-40]   [5-41]

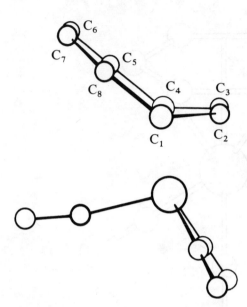

Fig. 5-7. X-ray diffraction data for cyclo-octatetraene-iron tricarbonyl. Bond distances are $C_1—C_2 = C_3—C_4 = 1.42$, $C_2—C_3 = 1.42$, $C_4—C_5 = C_1—C_8 = 1.45$, $C_5—C_6 = C_7—C_8 = 1\cdot34$, $C_6—C_7 = 1.49$, $Fe—C_1 = Fe—C_4 = 2.18$, $Fe—C_2 = Fe—C_3 = 2\cdot05$, Fe—C (carbonyl) $= 1.80$ (av), and C—O $= 1.13$ (av), all $\pm$ about 0.02 Å. Bond angles are $C_1—C_2—C_3 = C_2—C_3—C_4 = 124.6°$, $C_3—C_4—C_5 = C_2—C_1—C_8 = 132.4°$, $C_4—C_5—C_6 = C_1—C_8—C_7 = 133.2$, and $C_5—C_6—C_7 = C_8—C_7—C_6 = 131.8°$, all $\pm$ about 1°, and to be compared with 135° in the regular plane octagon. The angle between normals to the two planes in C_8H_8 is 41° in $C_8H_8Fe(CO)_3$. Reprinted with permission from B. Dickens and W. N. Lipscomb, *J. Am. Chem. Soc.* **83**:489 (1961).

tentatively proposed. The cyclo-octatetraene ring in this structure maintains a rather planar configuration. Although [5-41] satisfies the experimental data, x-ray diffraction analysis revealed an unexpected structure (Fig. 5-7).

The iron tricarbonyl group is attached to a "butadiene unit" of the cyclo-octatetraene ligand. This structure was simply ignored by early investigators.

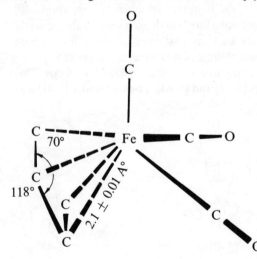

Fig. 5-8. Structure of butadieneiron tricarbonyl. Reprinted with permission from O. S. Mills and G. Robinson, *Acta Cryst.* **16**:758 (1963).

At present it is understood that cyclo-octatetraene iron tricarbonyl may have a different structure in solution than it does in the crystalline state. Indeed, different protons are detected at different temperatures, which has led to such descriptions as ring whizzing, dynamic equilibrium, and stereochemical nonrigidity in an attempt to describe the bonding.

An analogous structure is illustrated by the butadieneiron tricarbonyl complex (Fig. 5-8). The iron atom is approximately equidistant from the four carbon atoms of the butadiene system.

5-3. ELUCIDATION OF STRUCTURE WITHOUT CONFIRMATION BY X-RAY DIFFRACTION DATA

The following two examples, which do not involve confirmation by x-ray diffraction analysis, provide further interesting aspects of the structure elucidation of metal π-complexes.

Di-Diphenylchromium Iodides

An orange chromium compound was isolated from the reaction mixture of phenylmagnesium bromide and anhydrous chromium(III) chloride in ether. Treatment of this compound with KOH and KI gave golden orange chromium derivatives. These compounds were analyzed as $C_{12}H_{10}CrI$ [5-42] and [5-43], $C_{18}H_{15}CrI$ [5-44] and [5-45], and $C_{24}H_{20}CrI$ [5-46]–[5-48]. Many structures were tentatively proposed, some of which are shown in [5-42]–[5-48].

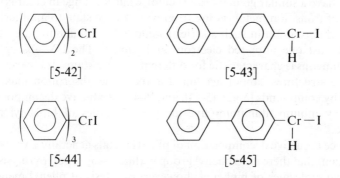

[5-42] [5-43]

[5-44] [5-45]

Several properties reported did not agree with the structures formulated above. These properties include: color, golden orange, and magnetic susceptibility, BM, 1.75.

The magnetic susceptibility determined for the series of compounds tri-, tetra-, and pentaphenylchromium iodide was found to be approximately

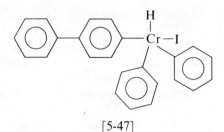

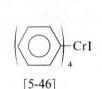

[5-46]   [5-47]

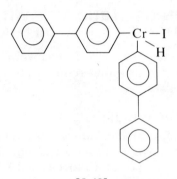

[5-48]

1.7 BM (indicates one free electron) for each compound. In addition, all compounds have a similar golden orange color while a change in color is characteristic of transition metals going from one valence state to another. Surely σ-bonded chromium in such a series should exhibit different valence states and not just one unpaired electron, as is noted. The UV absorption by chromium ions is characteristic for different valence states. It is clear that the tentative structures cannot account for the same absorption maxima exhibited by compounds [5-42], [5-44], and [5-46]. Reductive decomposition of [5-46] by lithium aluminum hydride gave 2 moles of biphenyl and no benzene.

Since reductive decomposition of phenyl metals generally gives benzene, it is evident that there is no phenyl group within [5-46]. Upon pyrolysis, [5-46] gives two molecules of biphenyl; however, pyrolysis of phenyl metals normally yields biphenyl quantitatively. Therefore, pyrolysis data cannot provide positive evidence of the existence of biphenyl in [5-46]. This case is different than for the tentative assignment of the structure of dibenzenechromium, where pyrolysis data is sufficient to suggest a π-complex bond between benzene and the chromium metal.

Considering these results along with the abnormal properties previously described, the structure of tetraphenylchromium iodide was reformulated as a π-complex [5-49].

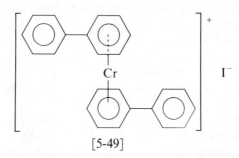

[5-49]

Reformulated structures were proposed for [5-42] and [5-44]. These structures are strongly supported by the valence bond and EAN theories.

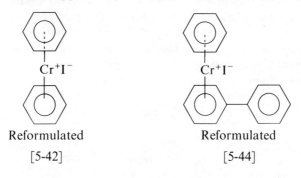

Reformulated
[5-42]

Reformulated
[5-44]

π-Cyclopentadienylnickel Azobenzene Complex

The reaction of di-π-cyclopentadienylnickel and azobenzene gives a purple–blue crystalline nickel complex, mp 118–119°. Based on the following data, structure [5-50] was assigned to the compound. Elemental analysis

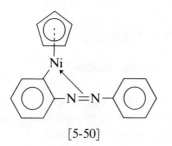

[5-50]

found: C, 66.7%; H, 4.67%; Ni, 19.1%; mol wt, 310; NMR, −5.33 (singlet), −6.92 (triplet), −7.93 (quadruplet), and −8.15 (doublet); and ppm with a ratio of 4.8:2.2:6.7:1, respectively. Treatment of the complex with $LiAlH_4$ in ether gave azobenzene. Treatment of the complex with $LiAlD_4$ in ether gave azobenzene that contains approximately 10% deuterium. UV λ_{max} was 252, 268, 345, and 585 mμ with respective molar extinction coefficients of 14,000, 13,050, 7550, and 6200 in 95% ethanol. (UV λ_{max} of azobenzene was 230, 318, and 440 mμ with molar extinction coefficients of 8370, 18,500 and 587.) The absorption in the 318 mμ region is attributed to the conjugation of the unsaturated nitrogen with the phenyl rings.

π-Complex bonding between the cyclopentadienyl ring and the nickel atom is assigned from the NMR data 5.33 τ (singlet). The π-complex bond between the azobenzene and the nickel metal atom is assigned from the ultraviolet data. Absorption around 318 mμ, which is attributed to the conjugation of the unsaturated nitrogen with the phenyl ring, is not exhibited in the absorption spectrum of the complex. A σ-bond between the phenyl group in azobenzene and the nickel atom is assigned from the results of the reductive decomposition of the complex with lithium aluminum deuteride. The position of the σ-bond is not established since the position of the deuterium atoms in deutero azobenzene has not been determined.

The participation of the π-coordination of the N=N linkage to the metal is not ambiguously proved as yet. σ-Coordination using a nitrogen lone pair as shown in structure [5-51] cannot be excluded as a possibility.

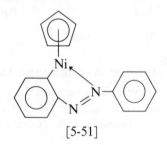

[5-51]

5-4. NOVEL π-COMPLEXES

Carborane–Metal Complexes

Transition metal complexes of boron hydrides and carboranes also have been reported recently. The number of papers concerned with this fascinating new area of inorganic chemistry is steadily increasing, and the topic could develop into a distinct research area comparable to coordination and organo-

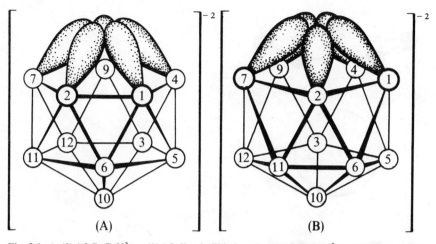

Fig. 5-9. *A* : (3)-1,2-$B_9C_2H_{11}^{2-}$ π-(3)-1,2-dicarbollide ion. *B* : (3)-1,7-$B_9C_2H_{11}^{2-}$ π-(3)-1,7-dicarbollide ion. $\bigcirc$ = B—H; $\bullet$ = C—H; and ③ = open position left by removal of B-atom 3. For details of nomenclature see *Inorg. Chem.* **2** : 1087 (1963).

metallic chemistry. While the reported boron hydride complexes display σ-bonding modes, a rather extensive number of transition metal compounds of carborane anions have been prepared and formulated as π-bonded species. In several cases single-crystal x-ray studies have confirmed their formulation as metallocene analogues. π-Complexes of iron, cobalt, nickel, palladium, manganese, rhenium, tungsten, and molybdenum and the two dianions shown in Fig. 5-9 have been isolated and characterized.

The sandwich-type bonding of the dicarbollyl group has been established by a single-crystal x-ray diffraction study for the compound π-cyclopentadienyl-π-(3)-1,2-dicarbollyliron(III). This compound, prepared by the reaction of an equimolar mixture of sodium cyclopentadienide and disodium (3)-1,2-dicarbollide with iron(III) chloride, followed by air oxidation, has the structure shown in Fig. 5-10.

The distances between the iron and the atoms of the pentagonal face of the dicarbollyl group and the cyclopentadienyl ring carbon atoms are 2.07 Å. An ESR study of this paramagnetic (d^5) derivative shows the presence of a single unpaired electron that can best be described as occupying an e_{2g}^* orbital in the ferrocene MO model. While the $'$H NMR could not be obtained because of the paramagnetism of the compound, the $''$B NMR spectrum was measured. Because of the large chemical shifts and the singlet nature of the signals, the paramagnetic broadening of the $''$B resonances did not prevent the observation of a useful spectrum. An interesting but unexplained feature of the spectrum (Fig. 5-11) is the absence of $''$B-H spin–spin coupling.

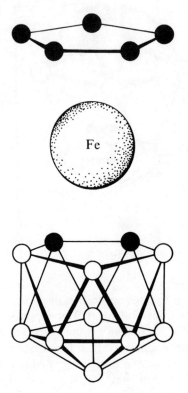

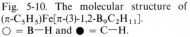

Fig. 5-10. The molecular structure of $(\pi\text{-}C_5H_5)Fe[\pi\text{-}(3)\text{-}1,2\text{-}B_9C_2H_{11}]$. ◯ = B—H and ● = C—H.

A sandwich structure for the dicarbollyl group would be expected to give a ^{11}B NMR spectrum having the intensity ratios $1:1:1:2:2:2$. From an examination of the spectrum it is concluded that the missing sixth peak (of relative intensity 1) lies under the b and c resonances. The high field resonance c of relative intensity 2 is assigned to the two borons bonded to the carbon atoms in the open face of the (3)-1,2-dicarbollyl group, since these would be expected to be greatly influenced by the unpaired electron on the iron. The a, b, c, and d resonances have not been assigned. The measurement of this spectrum, however, in conjunction with the definative x-ray study, allows the use of ^{11}B NMR spectroscopy as a "fingerprint" for π-bonding in other such compounds.

These bis-π-dicarbollyl derivatives of Fe(II) and Fe(III) also have been prepared, as shown in equation (5-2).

$$2[(3)\text{-}1,2\text{-}B_9C_2H_{11}]^{2-} + FeCl_2 \xrightarrow[N_2]{THF} [\pi\text{-}(3)\text{-}1,2\text{-}B_9C_2H_{11}]_2Fe^{2-} + 2Cl^- \tag{5-2}$$

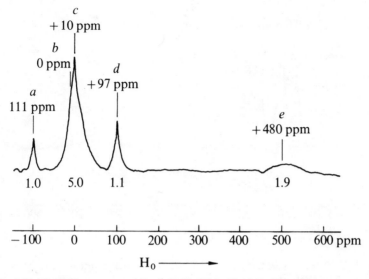

Fig. 5-11. "B NMR spectrum of $[(\pi\text{-}C_5H_5)Fe(III)(\pi\text{-}(3)\text{-}1,2\text{-}B_9C_2H_{11})]$ 19.3 Mc/sec., chemical shifts are relative to $BF_3 \cdot O(C_2H_5)_2 = 0$.

The Fe(II) derivative is not obtained directly but rather by the sodium amalgam reduction of the Fe(III) compound in acetone or acetonitrile. The latter is isolated as the tetramethylammonium derivative from aqueous solution following air oxidation of the initial reaction solution after stirring under nitrogen for an hour. The light red iron(III) compound, an analogue of the ferricenium ion, is quite stable toward strong acids and is thermally stable up to 300°. The ferrous derivative, while quite air sensitive in solution, is oxidized only slowly in the solid state and shows the same thermal stability of the oxidized compound. This greater stability of the (3)-1,2-dicarbollyl complexes of iron compared to ferrocene and the ferricinium ion may be the result of their more favorable charges as well as the fact that the π-orbitals in the $B_9C_2H_{11}^{2-}$ ligands are directed toward the iron atom, resulting in increased overlap with the iron d orbitals compared to the cyclopentadienyl analogs.

The iron(II) compound is readily reconverted to the ferric state by the readmission of air. The reversibility of this redox reaction has been demonstrated by polarography in acetone solution.

These compounds have been characterized by elemental analysis, IR, electronic spectra 'H NMR, and magnetic susceptibility measurements. The similarity of the "B NMR (Fig. 5-12) with that of $(\pi\text{-}C_5H_5)$ Fe(III) $(\pi\text{-}B_9C_2H_{11})$ has been taken as evidence of a sandwich structure for the

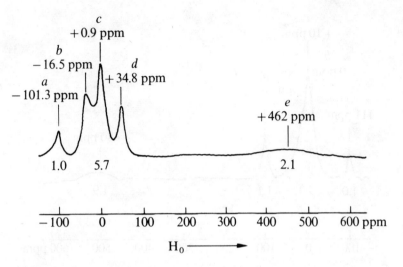

Fig. 5-12. ^{11}B NMR spectrum of $(CH_3)_4N[(B_9C_2H_{11})_2Fe(III)]$; chemical shifts are recorded relative to $BF_3 \cdot 0(C_2H_5)_2 = 0$ at 19.3 Mc/sec.

$B_9C_2H_{11}^{2-}$ groups. The simplicity of the spectrum also suggests that the two ligands are equivalent and in rapid rotation with respect to one another on the NMR time scale.

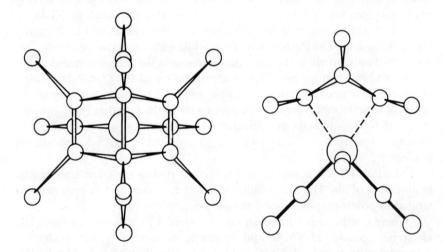

Fig. 5-13. Structure of hexamethyl bicyclo [2.2.0] hexa-2,5-diene chromiumtetracarbonyl.

Hexamethyl Bicyclo [2.2.0] Hexa-2,5-Diene Chromiumtetracarbonyl

Chromium hexacarbonyl generally reacts with arenes and is stabilized as the tricarbonyl complex. Hexamethyl bicyclo [2.2.0] hexa-2,5-diene chromiumtetracarbonyl is an unusual example of a stabilized chromiumtetracarbonyl. The x-ray pattern reveals a C_{2v} symmetry (Fig. 5-13). This particular bicyclo [2.2.0] diene system may effect the stabilization of the electron configuration of the complex.

5-5. EXERCISES

5-1.   Show the structures of possible isomers:

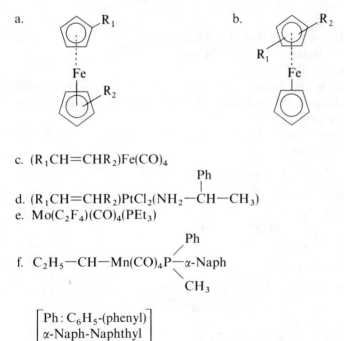

c. $(R_1CH=CHR_2)Fe(CO)_4$

$$Ph$$
d. $(R_1CH=CHR_2)PtCl_2(NH_2-\overset{|}{C}H-CH_3)$

e. $Mo(C_2F_4)(CO)_4(PEt_3)$

f. $C_2H_5-CH-Mn(CO)_4P\overset{\diagup Ph}{\underset{\diagdown CH_3}{-}}\alpha\text{-Naph}$

$$\left[\begin{array}{l}Ph: C_6H_5\text{-(phenyl)}\\ \alpha\text{-Naph-Naphthyl}\end{array}\right]$$

5-2.   What stereochemistry would one predict for a four-coordinate complex on the basis of crystal field theory? molecular orbital theory?

5-3.   What physical tool could be employed to determine the configuration of $[PdCl_2(C_6H_5CN)_2]$ as *cis* or *trans* (square planar complex).

5-4.   Both the structures shown follow the EAN rule. What two physical structural tools could be employed to determine the true structure?

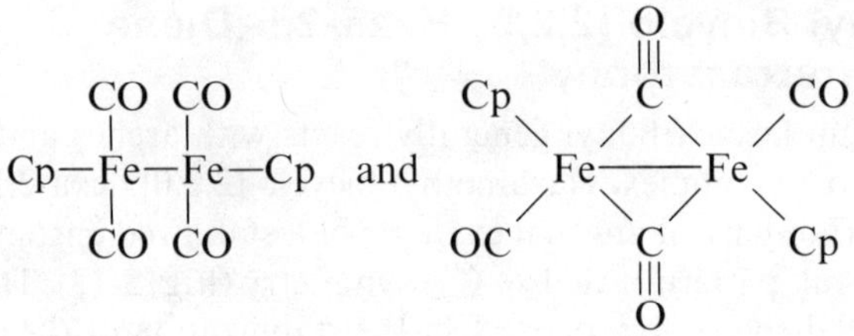

5-5. Indicate the possible methods for the determination of the structure for the following:

a. $(CH_2{=}CH{-}CH{=}CH_2)Fe(CO)_3$

b. 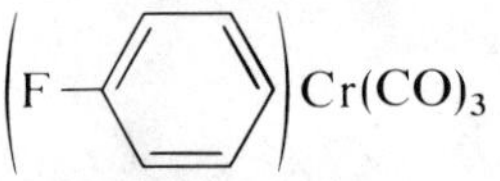

c. $CH_3{-}CH{=}CH{-}CH_2{-}Mn(CO)_5$

d. $(\pi\text{-}C_5H_5)Mn(CO)_2{\cdot}PhC{\equiv}CPh$

e. $PhC{\equiv}CPhCo_2(CO)_8$

5-6. Complete the following equations and describe how to verify the structures of the respective products (without the use of x-ray diffraction analysis):

a. pentacarbonyl (triphenylstannyl) manganese + tetraphenylcyclopentadicne

b. 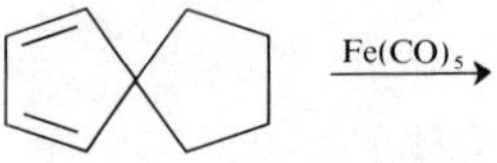

c. 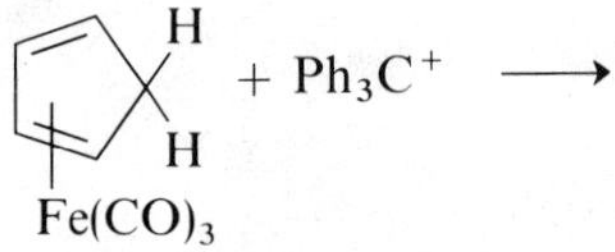

d. 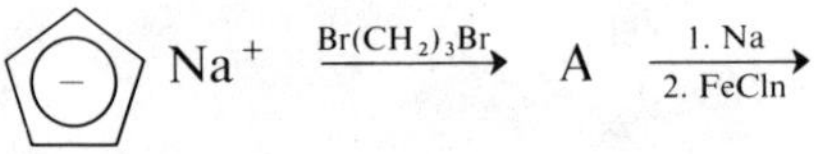

e. 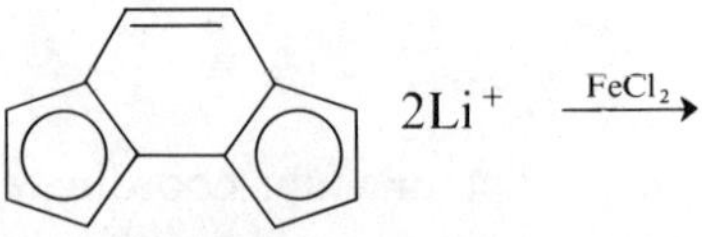

f. 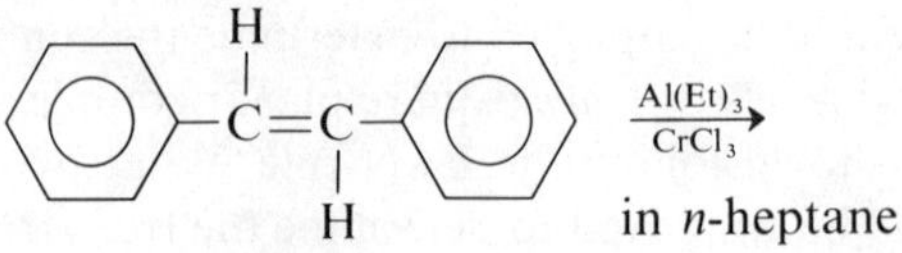

5-7. Excess iron nonacarbonyl$Fe_2(CO)_9$ was stirred with 3-chloro-(2-chloromethyl) propene in ether at room temperature for 12 hr. The pale yellow complex, mp 28–29°, was isolated from the reaction mixture. Propose a plausible structure from the following data:

Analysis: C, 43.29 and H, 2.90; mass spectrum m/e 194, 166, 138, and 110; IR 1998 and 2064 cm^{-1}; NMR 8.00τ(s); π-cyclobutadiene iron tricarbonyl 6.09τ; π-allyl-π-cyclopentadienyl iron monocarbonyl; 7.33τ (syn), and 9.32τ (anti).

5-6. BIBLIOGRAPHY

F. Basolo and R. Johnson, "Coordination Chemistry," W. A. Benjamin, New York (1964).

G. E. Coates, "Organometallic Compounds," 2nd ed., Wiley–Methuen, London (1960).

F. A. Cotton and G. Wilkinson, "Advanced Inorganic Chemistry," John Wiley & Sons, New York (1966).

E. O. Fischer and H. Werner, "Metal π-Complexes," Vol. 1, Elsevier Publishing, Amsterdam (1966).

E. M. Larsen, "Transition Elements," W. A. Benjamin, New York (1964).

H. Zeiss, P. J. Wheatley, and H. J. S. Winkler, "Benzenoid-Metal Complexes," Ronald Press, New York (1966).

BIBLIOGRAPHY

CHAPTER 6

Reactions of Metal π-Complexes

Metal π-complexes are susceptible to a wide range of chemical reagents. However, the three major groups of metal π-complexes—π-olefin, π-cyclopentadienyl, and π-arene metal—demonstrate distinctly characteristic reactions. π-Cyclopentadienyl complexes (metallocenes) exhibit a high degree of aromaticity and undergo many typical aromatic substitution reactions. The π-arene complexes, on the other hand, do not exhibit a discernible degree of aromaticity. This comparative behavior invites diverse speculation on the exact nature of the π-complex bond. The reactions of most π-olefin complexes often parallel those of uncomplexed olefins. Differing behavior is generally explained on the basis of the strength and stability of the metal–olefin bond, which resists attack.

Although most physical properties, and particularly the structure of metal π-complexes, are logically interpreted by the application of the basic principles of coordination chemistry, these established principles do not suitably explain reaction anomalies of the different groups of metal π-complexes. This chapter provides a description of the reactions characteristic of each major group and also of reactions that are common to all.

6-1. REACTIONS OF CYCLOPENTADIENYL π-COMPLEXES

As indicated, the most significant feature of the reactions of π-cyclopentadienyl complexes in general and ferrocene in particular involves their

119

aromatic nature. The resonance stabilization energy calculated for ferrocene is 50 kcal/mole, and in that sense it is more aromatic than benzene. At this writing there is no completely adequate description of the bonding in ferrocene. At one extreme we find descriptions of ferrocene possessing a covalent ring-to-metal bond, as described in Chapter 1. Thus a covalent bond would have little charge displacement and would leave the metal and rings essentially neutral. The absence of appreciable charge separation is supported by the near identity in the acidity constants of ferrocenic and benzoic acids. At the other extreme, ferrocene is seen as consisting of cyclopentadienide anions and an Fe(II) ion forming an octahedral coordination compound.

Reactions of Ferrocene

Ferrocene undergoes a large number of typical ionic aromatic substitution reactions that include Friedel–Crafts acylation, alkylation, metalation, sulfonation, and aminomethylation. Figure 6-1 outlines some of the more typical reactions of ferrocene.

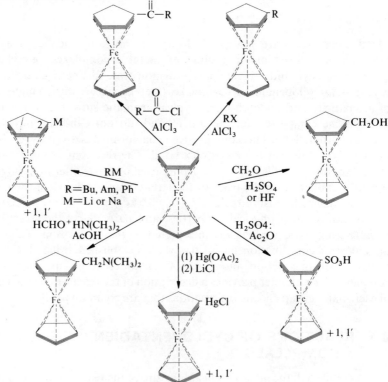

Fig. 6-1. Ionic aromatic substitution of ferrocene.

Friedel–Crafts Acylation

The acylation of metallocenes proceeds rather easily. Thus this reaction has been most extensively studied. The equimolar reaction of ferrocene and acetyl chloride in the presence of aluminum chloride gives monoacetylferrocene almost exclusively. When an excess of acetyl chloride and aluminum chloride is used, a mixture of two isomeric diacetylferrocenes is produced. The heteroanular disubstituted derivative, 1,1′-diacetylferrocene [6-1],* and the homoanular isomer, 1,2-diacetylferrocene [6-2], are obtained in a ratio of

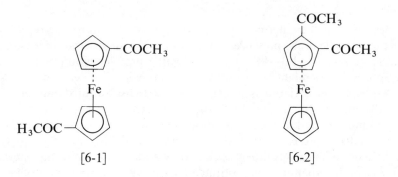

[6-1]   [6-2]

60:1. The first acetyl group appears to deactivate the π-cyclopentadienyl ligand toward further electrophilic substitution. Thus, the second acetyl group enters the other ring.

The high reactivity of ferrocene in acetylation has been confirmed in a competitive reaction between ferrocene and anisole. When ferrocene and anisole were allowed to compete for a limited amount of acetyl chloride and aluminum chloride, acetylferrocene was formed to the exclusion of any detectable amount of methoxyacetophenone. Furthermore, ferrocene undergoes acetylation with acetic anhydride and stannic chloride or boron trifluoride even under mild conditions.

The effects of the strength of π-complex bonding are observed in a comparison of the reactivity among ferrocene, ruthenocene, and osmocene. The electrophilic reactivity of these metallocenes decreases in the following order: ferrocene > ruthenocene > osmocene (Table 6-1).

The decreased electrophilic reactivity in ruthenocene and osmocene shows that the cyclopentadienyl rings in ruthenocene and osmocene are bound more tightly to the central metal atom than in ferrocene. Therefore, the π-electron density around the ring is decreased. When monosubstituted ferrocene is acylated, the nature of the substituent markedly affects the

---

* Double numbers in brackets refer to structural formulas, which sometimes appear within figures or equations.

Table 6-1. Friedel–Crafts Reaction of Metallocenes with Acetylchloride

Metallocene	Reaction time, hr	Yield of mono- and diketone, %
Ferrocene	2.25	69
Ruthenocene	2.75	53
Osmocene	2.25	23

orientation of the entering acetyl group. Monoethyl ferrocene reacts with acetyl chloride in the presence of aluminum chloride to yield three isomeric acetyl ferrocenes: 1,3-, 1,2-, and 1,1′-, with a product ratio of 4.2:1.4:1.0, respectively. As might be expected by the presence of an electron-releasing alkyl group, the acylation is more facil in the ethyl-substituted ring compared to the unsubstituted ring. In ferrocene, electrophilic acylation of the 3-position is favored over the 2-position; in benzene, the 2- or 4-position can be acylated more easily than the 3-position. This is a unique property of ferrocene. When a substituent is electron withdrawing, for example, cyano-ferrocene, heteroannularly substituted 1,1′-acetylcyanoferrocene is obtained.

Friedel–Crafts Alkylation

1,1-Diferrocenylethane was obtained as the anomalous product of the Friedel–Crafts reaction of ferrocene and ethylene dichloride, equation (6-1).

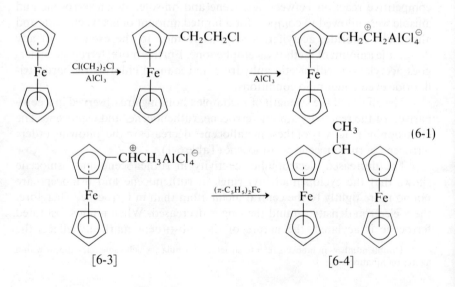

$$(6-1)$$

This reaction is unusual in comparison to the reaction of benzene and 1,2-dichloroethane, which gives dibenzyl. Therefore, the formation of 1,1-diferrocenylethane distinctly shows that a hydride ion rearrangement takes place in going from [6-3] to [6-4]. The rearrangement might result from the remarkable stability of the α-ferrocenyl carbonium ion.

Aminomethylation

Ferrocene undergoes a Mannich-type reaction with formaldehyde and dimethylamine to form dimethylaminomethyl-ferrocene, which is readily converted to a methiodide equation (6-2). This quarternary ammonium salt

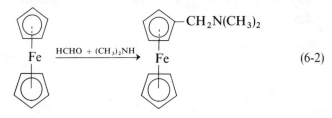

$$(6-2)$$

is an important intermediate in the synthesis of ferrocene derivatives, such as methyl- and dimethylaminoethyl-ferrocene, ferrocenylacetonitrile, and ferrocenylcarbinol (see Fig. 6-4).

Sulfonation

Ferrocene can be readily sulfonated by sulfuric acid or chlorosulfonic acid in acetic anhydride to form ferrocene-sulfonic acid and heteroannular disulfonic acid. π-Cyclopentadienylrhenium tricarbonyl can be sulfonated with concentrated sulfuric acid in acetic anhydride, the product being isolated as the p-toluidine salt—showing the aromatic nature of the π-C_5H_5 ring in this complex.

Formylation

Ferrocene is readily formylated with N-methylformanilide in the presence of phosphorous oxychloride. This reaction is also characteristic of highly reactive aromatic rings.

Condensation

Ferrocene condenses with formaldehyde in either concentrated sulfuric acid or liquid hydrogenfluride, followed by reduction, to give a compound containing two ferrocene nuclei and two methylene groups. The compound obtained has been identified as 1,2-diferrocenyl ethane. On the other hand, this condensation using excess ferrocene primarily gives diferrocenylmethane.

The mechanism of these reactions presumably involves the initial formation of ferrocenylcarbinol, followed by ionization in the strongly acidic media to the relatively stable ferrocenylmethylcarbonium ion. The latter is assumed to be in equilibrium with a ferricinium ion radical, which will dimerize to the dication, or react with ferrocene to form a diferrocenylmethane cation, equation (6-3).

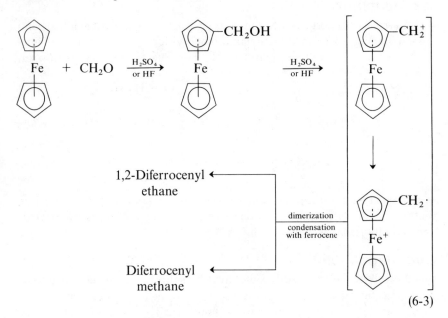

$$(6\text{-}3)$$

This mechanism is supported by the fact that ferrocenylcarbinol reacts with ferrocene under the same conditions to form diferrocenylmethane.

Arylation

The most significant radical substitution reaction of ferrocene is noted upon reaction with aryl diazonium salts giving an arylation product, equation (6-4).

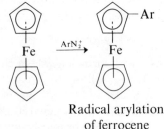

$$(6\text{-}4)$$

Radical arylation
of ferrocene

A plausible mechanism, equation (6-5), involves initially the gain of an electron by the diazonium cation to form a diazo radical. This subsequently decomposes to give N_2 and an aryl radical, equation (6-6), which reacts with ferrocene, equation (6-7), leading to the final product, equation (6-8). Later

$$(\pi\text{-}C_5H_5)_2Fe + ArN_2^+ \longrightarrow (\pi\text{-}C_5H_5)Fe^+ + ArN_2\cdot \qquad (6\text{-}5)$$

$$ArN_2\cdot \longrightarrow Ar\cdot + N_2 \qquad (6\text{-}6)$$

$$Ar\cdot + (\pi\text{-}C_5H_5)_2Fe \longrightarrow (\pi\text{-}ArC_5H_4)Fe(\pi\text{-}C_5H_5) + H\cdot \qquad (6\text{-}7)$$

$$H\cdot + (\pi\text{-}C_5H_5)_2Fe^+ \xrightarrow{\;i\;} H^+ + (\pi\text{-}C_5H_5)_2Fe \qquad (6\text{-}8)$$

an alternative mechanism, an arylation mechanism involving the internal rearrangement of a ferrocenediazonium salt charge-transfer complex as an intermediate, equation (6-9), was proposed. It is not yet clear which of these mechanisms is the correct one.

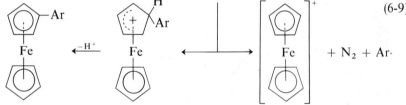

charge transfer complex

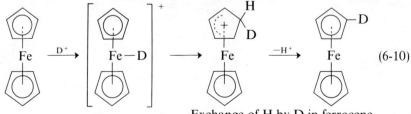

Exchange Reaction

Ferrocene undergoes a hydrogen–deuterium exchange reaction only in an acid media such as deuterated hydrochloric acid and deuterated trifluoroacetic acid. This may be because of the possibility that the deuterium cation can associate with the metal nucleus followed by a rearrangement and substitution on the ring in the acid media, equation (6-10).

Exchange of H by D in ferrocene

Reduction

Ferrocene does not easily undergo reduction. Under the extreme condition of 280 atm of hydrogen at 300–340°C, using Raney nickel, more than half the ferrocene remains unchanged. Ferrocene can be reductively decomposed in a Li-liquid NH_3 system.

Reactions of Ferrocene Derivatives

We can grasp the extent to which the chemistry of the metallocenes has been studied by illustrating only a limited number of typical reactions of various ferrocene derivatives (Figs. 6-2 and 6-3).

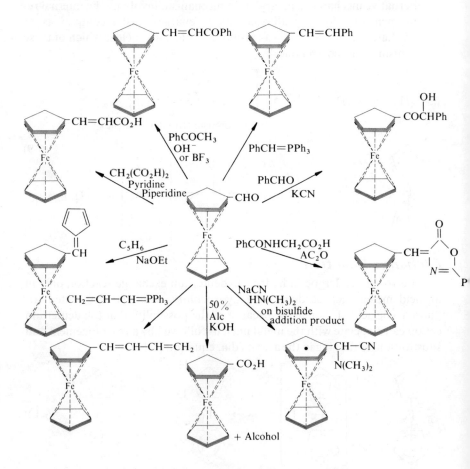

Fig. 6-2. Typical reactions of ferrocenecarboxaldehyde.

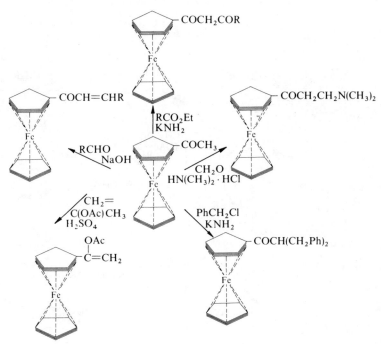

Fig. 6-3. Typical reactions of acetylferrocene.

Aminomethylation of ferrocene is significant because of the large number of ferrocene derivatives that can be prepared from the quarternary ammonium salt, some of which are illustrated in Fig. 6-4.

One of the most interesting reactions of ferrocene derivatives is hetero-anular bridging by the cyclization of a ω-ferrocenyl aliphatic acid, equation (6-11). If $n = 2$, the homoanularly cyclized product is obtained; if $n = 3$ or 4,

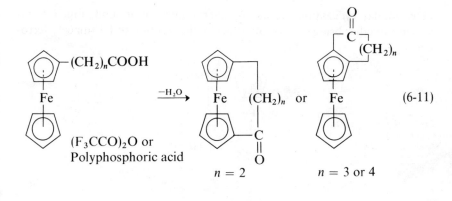

$$\hspace{10cm} (6\text{-}11)$$

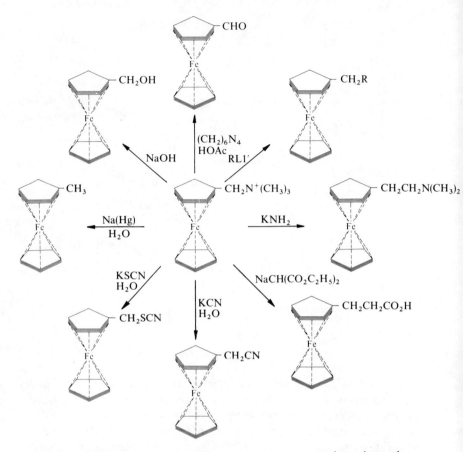

Fig. 6-4. Representative reactions of ferrocene quaternary ammonium salt complex.

heteroanular cyclization occurs. A heteroanular compound, Fig. 6-5, involving an ether bridge can be obtained by the reaction of 1,1'-diacetylferrocene.

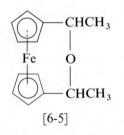

[6-5]

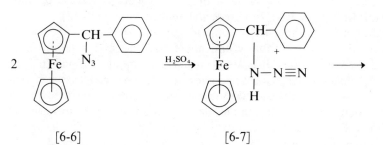

$$2 \qquad \xrightarrow{H_2SO_4} \qquad \longrightarrow$$

[6-6]   [6-7]

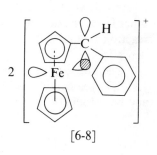

[6-8]

(6-12)

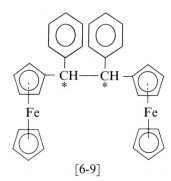   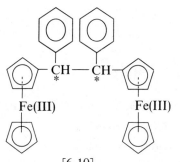

[6-9]   [6-10]

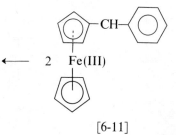

[6-11]

Ferrocenylphenylcarbinyl azide undergoes an acid-catalyzed decomposition to form 1,2-diphenyl-1,2-diferrocenylethane, equation (6-12).

The formation of the diasteroisomeric structure [6-11] is another manifestation of the unusual stability and, therefore, ease of formation of α-ferrocenyl carbonium ions. It is assumed that the conjugated acid [6-7] of ferrocenylphenylcarbinyl azide undergoes an anchimerically assisted displacement of hydrogen azide to give the ferrocenylphenylmethyl cation [6-8]. The latter then undergoes intramolecular oxidation reduction to form the radical ion [6-9]. This undergoes subsequent coupling to form the diastereoisomeric dipositive cation [6-10]. In part these cations are reduced to the diastereoisomeric form [6-11] by intermolecular oxidation reduction with other ferrocene derivatives present in the reaction mixture. This sequence of reactions is analogous to a similar sequence described in the section on condensation.

The cleavage of ferrocene by aluminum chloride in benzene produces 1,1'(-1,3-cyclopentylene) ferrocene [6-12]. Phenylcyclopentylferrocene is another product of this cleavage [6-13].

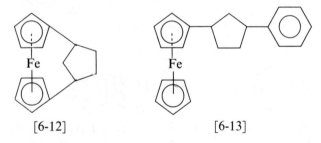

[6-12]   [6-13]

Reactions of Other Metallocenes
Addition Reactions

Tetrafluorethylene reacts with nickellocene to form [6-14], equation (6-13). On the other hand, cobaltocene undergoes addition of tetrafluoroethylene to yield the binuclear compound [6-15], equation (6-14).

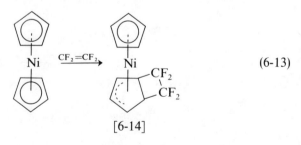

(6-13)

[6-14]

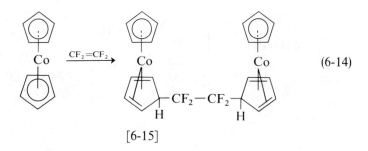

$$(6\text{-}14)$$

$$[6\text{-}15]$$

Some metallocenes show interesting behavior upon the addition of organic halides. For example, cobaltocene reacts with organic halides to form equimolar amounts of cobalticinium halide and the 1-endo addition compound, equation (6-15).

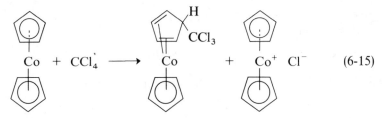

$$(6\text{-}15)$$

Generally the uncatalyzed addition of a weak acid to olefins does not readily occur. However, it is remarkable to find acetic acid adding smoothly to the double bond of vinylmetallocene. The relative rates of addition are shown in Table 6-2.

The greater rates of addition of acetic acid to vinylmetallocenes might result from the unusual stability of the carbonium ion, which is formed by the overlap of the *hag* molecular orbital of the metallocene nucleus with the available *p*-orbital of the positively charged α-carbon atom, equation (6-16).

The relative rates of addition can be attributed to the increasing overlap as the size of the metal atom in the metallocene increases.

Table 6-2. Relative Rate of Addition of Absolute Acetic Acid to Vinyl Metallocenes

Compound	Relative rate
Vinylferrocene	1.00
Vinylruthenocene	1.19
Vinylosmocene	4.62

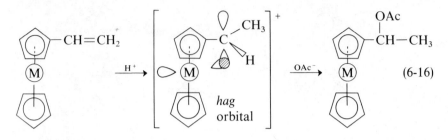

$$(6\text{-}16)$$

Reduction Reactions

The reduction of nickelocene with sodium amalgam in ethanol results in the partial reduction of one of the cyclopentadienyl rings, yielding π-cyclopentenyl-π-cyclopentadienylnickel, equation (6-17).

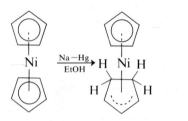

$$(6\text{-}17)$$

On the other hand, the reduction of cobalticinium ion or rhodenocinium ion with sodium borohydride or lithium aluminum hydride gives cyclopentadienylcobalt-cyclopentadiene and π-cyclopentadienyl rhodium-cyclopentadiene, respectively.

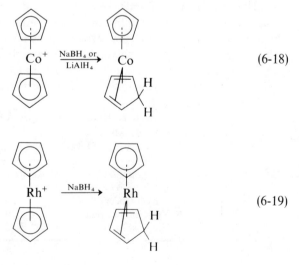

$$(6\text{-}18)$$

$$(6\text{-}19)$$

It is characteristic of these Co and Rh metallocenes to be reduced with hydride reagents to a noble gas electronic configuration without undergoing further reduction, equation (6-18) and (6-19).

π-Cyclopentadienyl iron tricarbonyl cation is reduced by a hydride ion to form π-cyclopentadienyl iron dicarbonyl hydride, equation (6-20). In this case the π-cyclopentadienyl ring remains intact.

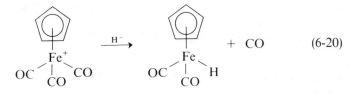

$$+ \quad CO \qquad (6\text{-}20)$$

Oxidation

The methyl substituent in cobalticenium ion can be oxidized to a carboxyl group using a relatively strong reagent such as $KMnO_4$. The cobalticenium ion is not attacked even by boiling aqua regia, equation (6-21).

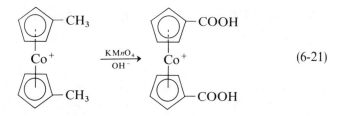

$$(6\text{-}21)$$

Solvolysis

Methylmetallocenylcarbinyl acetates are readily solvolyzed. In fact, the rates are greater than that of tritylacetate (Table 6-3). These rates of solvolysis also result from the stability of the corresponding α-metallocenyl carbonium ions.

Table 6-3. Relative Rates of Solvolysis of Methyl-metallocenylcarbinyl Acetates

Acetate	Relative rate (80% acetone, 30°)
Trityl	0.15
Methylferrocenylcarbinyl	1.00
Methylruthenocenylcarbinyl	1.36
Methylosmocenylcarbinyl	5.37

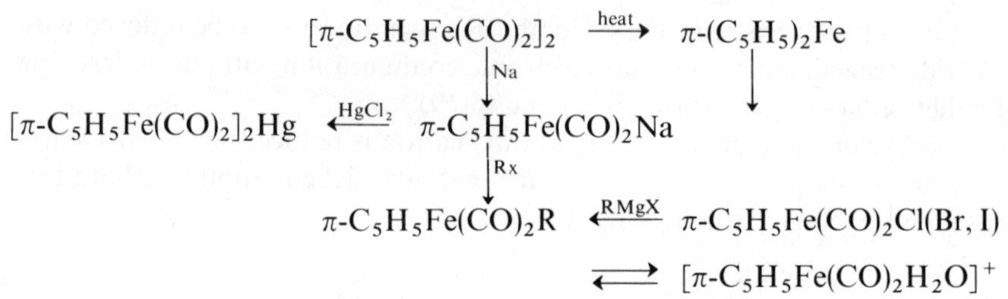

$$[\pi\text{-}C_5H_5Fe(CO)_2]_2 \xrightarrow{\text{heat}} \pi\text{-}(C_5H_5)_2Fe$$

$$\downarrow \text{Na}$$

$$[\pi\text{-}C_5H_5Fe(CO)_2]_2Hg \xleftarrow{\text{HgCl}_2} \pi\text{-}C_5H_5Fe(CO)_2Na$$

$$\downarrow \text{Rx}$$

$$\pi\text{-}C_5H_5Fe(CO)_2R \xleftarrow{\text{RMgX}} \pi\text{-}C_5H_5Fe(CO)_2Cl(Br, I)$$

$$\rightleftharpoons [\pi\text{-}C_5H_5Fe(CO)_2H_2O]^+$$

Fig. 6-5. Some reactions of mono-π-cyclopentadienyl compounds.

Reactions of Cyclopentadienyl Metal Carbonyls

These complexes also undergo many of the reactions shown for ferrocene, particularly the aromatic substitution reactions. An additional outline of some reactions of mono-π-cyclopentadienyl complexes are shown in Fig. 6-5. These compounds also undergo carbonyl substitution reactions.

6-2. REACTIONS OF OLEFIN π-COMPLEXES

Reactions involving olefin π-complexes may be directed at uncomplexed olefinic functions as well as complexed ones. Generally, reactions involving the former are not very different from those observed for free olefins. Reactions of the latter, however, are significantly altered by π-complex formation. Among the reactions of interest included are addition, elimination, and substitution. Typical organometallic rearrangements, for example, σ–π and π–σ (resulting from protonation or the elimination of a hydride ion by or from π-olefin complexes or by the addition of a hydride ion to the complexes), are described in other sections.

Addition Reactions

Hydrogenation

Two distinct observations are made with respect to the hydrogenation of olefin π-complexes with catalysts such as raney nickel. When the olefin is firmly bound to a metal, complete catalytic hydrogenation is prohibited. Thus π-cyclopentadienylcobalt cyclooctatetraene (6-22), π-cyclopentadienyl π-cyclopentadiene rhenium dicarbonyl (6-23), and π-cycloheptatriene iron tricarbonyl (6-24), each of which contains at least one uncomplexed double bond, readily undergo hydrogenation of the uncomplexed bonds only. The metal–olefin bond is preserved in each case.

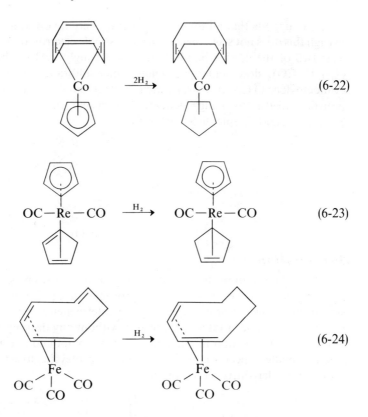

$$(6\text{-}22)$$

$$(6\text{-}23)$$

$$(6\text{-}24)$$

The second distinct observation involves hydrogenation of weakly bound olefin complexes in which hydrogenation is complete and the metal π-complex is destroyed. For example 1,5,9-cyclododecatriene-centro-nickel(0) absorbs hydrogen very readily, producing cyclododecane and metallic nickel. The ease of hydrogenation has led to the assignment of a zero valent nickel loosely associated with three double bonds, equation (6-25).

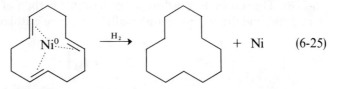

$$+ \ Ni \qquad (6\text{-}25)$$

An interesting application of this observation was encountered in the hydrogenation of cyclo-octatetraeneiron tricarbonyl $[C_8H_8Fe(CO)_3]$, which was not even partially hydrogenated by a platinum or nickel catalyst. This strongly suggests that all four double bonds were involved in bonding even though this does not seem reasonable on a simple electron basis. However, at least two of the double bonds cannot be strongly held to the metal since $C_8H_8Fe(CO)_3$ does undergo a Diels–Alder addition reaction with tetra-cyanoethylene (TCNE), a strong dienophile (6-26). The type of bonding has been described as involving valence tautomerism or ring whizzing (Chapter 4, Section 4-2, and Chapter 6, Section 6-6).

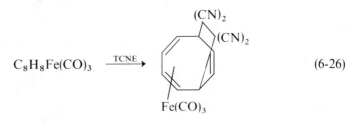

$$C_8H_8Fe(CO)_3 \xrightarrow{\text{TCNE}} \qquad\qquad\qquad (6\text{-}26)$$

Protonation

The addition of protons to some olefin π-complexes proceeds smoothly in some cases, while in other cases protonation can be accompanied by structural changes. For example, cyclo-octatetraeneiron tricarbonyl can be protonated in concentrated sulfuric acid without any decomposition. Upon dilution of the reaction mixture the original compound can be recovered. The protonated species is found to contain a bicyclic structure, a bicylo-5,1,0-octane derivative, equation (6-27A).

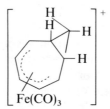

$$(6\text{-}27A)$$

Butadieneiron tricarbonyl also can be easily protonated, equation (6-27B). The proton is attached to the terminal carbon of the π-complexed butadiene, and the corresponding π-allyl derivative is isolated.

$$(6\text{-}27B)$$

Addition of water to the protonated complex gives methyl ethyl ketone since the addition occurs via a protonated species. In contrast to the behavior of the iron complex, π-cyclopentadienylcobalt cyclo-octatetraene cannot be protonated analogously. The protonated species is unstable and decomposes easily, equation (6-28).

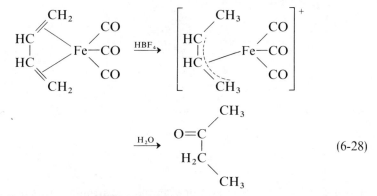

$$\xrightarrow{\text{H}_2\text{O}} \qquad (6\text{-}28)$$

Addition of Hydrogen Chloride

It has been reported that a diene iron carbonyl complex does not maintain its configuration upon the addition of hydrogen chloride, as evidenced by the different salts obtained from the reaction of butadieneiron tricarbonyl with HBF_4 or HCl followed by $AgBF_4$. Thus piperylene $Fe(CO)_3$ reacts with HCl to give the syn-syn dimethyl allyl derivative [6-16].

The NMR spectrum of the salt obtained from [6-16] upon reaction with silver perchlorate shows only one type of methyl group. This is not noted in the spectrum of the analogous salt protonated with HBF_4. Since inversion does not occur, the NMR spectrum reveals two types of methyl groups derived from this syn-anti derivative, equation (6-29).

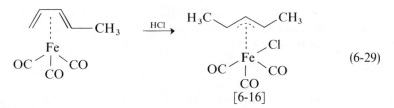

$$(6\text{-}29)$$

A mechanism involving *cis* addition of HCl and rotation to permit the halogen attack on the metal is probably involved, equation (6-30).

Elimination Reactions

The oxidation or reduction of the central metal atom in a π-complex occurs under varying reaction conditions and generally results in the formation of a more stable complex with altered geometry. An elimination reaction

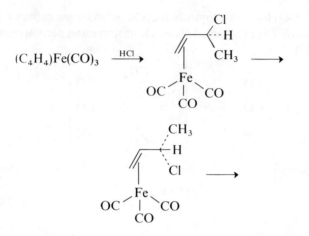

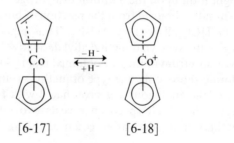

$$(6\text{-}30)$$

may accompany this process. For example, the aromatization of π-cyclopen-tadienyl π-cyclopentadiene cobalt [6-17] to π-dicyclopentadienyl cobalt hydride formally results in the elimination of hydride ion, equation (6-31).

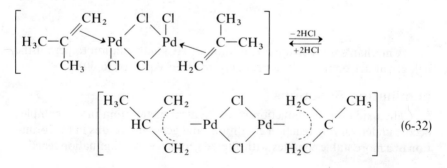

[6-17]   [6-18]

$$(6\text{-}31)$$

This reaction is reversible by the addition of hydride ion to [6-18]. Similar elimination reactions are observed for the palladium complex. An example of the use of hydrogen chloride in a reversible system is given in equation (6-32).

$$(6\text{-}32)$$

Substitution Reactions

Substitution is facilitated for a substituent attached to an α-carbon in π-allyl complexes, equation (6-33). This activation toward substitution has

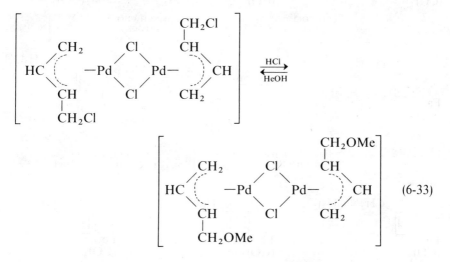

$$(6\text{-}33)$$

been likened to α-carbonium ion stabilization in ferrocene derivatives. Substitution on the carbon directly π-complexed to the metal has not been demonstrated. A stepwise substitution through a stabilized intermediate is shown in equation (6-34). Thus a substituted π-cycloheptatriene chromium-

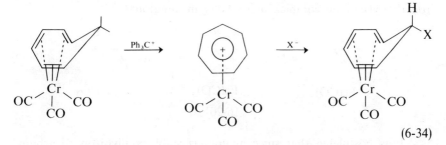

$$(6\text{-}34)$$

carbonyl complex is aided by the elimination of a hydride ion and formation of a tropylium π-complex intermediate.

An interesting recent development involves the suggested aromaticity of cyclobutadieneiron tricarbonyl [6-19]. This complex is aromatic in the sense that it undergoes electrophilic substitution reactions that parallel those of ferrocene. Figure 6-6 depicts the complete reaction schemes followed for electrophilic substitution and characterization of the reaction products.

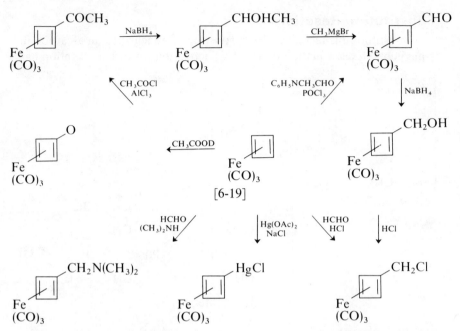

[6-19]

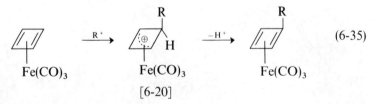

Fig. 6-6. Reactions demonstrating the "aromatic" nature of cyclobutadieneiron tricarbonyl toward electrophilic substitution.

The aromatic character of complex [6-19] is readily rationalized by the conventional mechanism for electrophilic substitution involving a π-ally-iron tricarbonyl cationic complex [6-20] shown in equation (6-35).

$$ (6\text{-}35) $$

[6-20]

It is postulated that since analogous stable π-allyl-iron tricarbonyl cationic salts have been isolated, [6-20] may be expected to provide a low-energy pathway for the substitution process.

6-3. REACTIONS OF ARENE METAL π-COMPLEXES

Generally, arene π-complexes do not undergo the reactions characteristic of benzene and its derivatives. However, arene π-Complexes do undergo a limited number of substitution, addition, expansion, and condensation reactions.

Substitution Reactions

The high reactivity of ferrocene toward electrophilic substitution was described earlier. Similar reactions of arene π-complexes result in the cleavage of the complexed bonds. For example, Friedel–Crafts acylation, metalation with butyllithium, mercuration, or nitration of ditoluenechromium, which is more soluble in organic solvents than the slightly soluble dibenzenechromium, failed to give any substitution products.

Acylation

Benzenechromium tricarbonyl, however, does undergo a Friedel–Crafts acylation. Furthermore, the acylation of toluenechromium tricarbonyl using acetyl chloride and aluminum bromide in carbon disulfide gives ortho, meta-, and para-substituted derivatives in the ratio of $39:15:46$. The orientation of substituent groups is much different from that of toluene itself, which mainly yields the p-substituted derivative (92%) and a small amount of the o-substituted derivative (80%).

Metalation

Dibenzenechromium(0) can be metalated with sodium and, in turn, converted to a methyl ester or a ketone, equation (6-36). Reaction of ditoluene chromium(0) with mercuric compounds does not give any mercurated derivatives but rather metallic mercury and an oxidized ditoluene chromium(I) cation. Ditoluenechromium undergoes metalation with butyllithium to give $(LiC_6H_4 \cdot CH_3)_2Cr$, which reacts further with carbon dioxide to yield $(CH_3 \cdot C_6H_4 \cdot COOLi)_2Cr$.

$$Cr(C_6H_6)_2 \xrightarrow[\text{Hexane}]{\text{AmNa}} Cr(C_6H_5Na)_2 + Cr(C_6H_6)(C_6H_5Na)$$

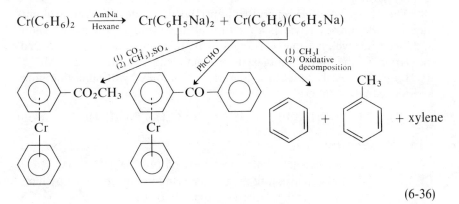

$$(6\text{-}36)$$

Addition Reactions

Some arene metal π-complex cations undergo nucleophilic attack by an anion, such as a hydride ion or a phenyl anion. In these reactions the arene is converted to a cyclohexadienyl anion. For example, the [dibenzenerhenium]$^+$ cation undergoes the addition of a hydride ion to give a new complex, $(C_6H_6)(\pi\text{-}C_6H_7)Re$, equation (6-37). The reaction of [dibenzenerhenium]$^{2+}$ with phenyllithium gives $(\pi\text{-}C_6H_5\cdot C_6H_6)_2Re(0)$, equation (6-38).

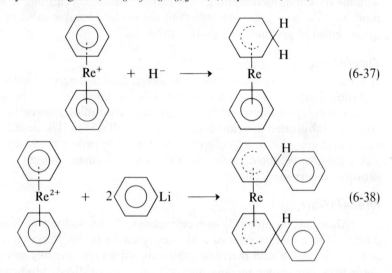

The following complexes also are obtained under similar conditions: $(\pi\text{-}C_6H_5\cdot C_6H_6)Mn^+(CO)_3$, $(C_6H_6)(\pi\text{-}C_6H_7)Re^+$, $(\pi\text{-}C_6H_5\cdot C_6H_6)$ $(\pi\text{-}C_5H_5)Fe^{2+}$, $(\pi\text{-}C_6H_5\cdot C_6H_6)_2Ru^{2+}$, $(C_6H_6)(\pi\text{-}C_6H_8)Ru(0)$, and $(\pi\text{-}C_6H_7)_2Ru^{2+}$.

The reaction of dibenzeneruthenium(II) perchlorate with lithium aluminum hydride, however, yields a mixture of di-π-cyclohexadienylruthenium and (benzene)(cyclohexadiene) ruthenium(0). If sodium borohydride is used, the latter is the only complexed product obtained. $[(\pi\text{-}C_5H_5)^-Mo^{2+}$-$(C_6H_6)]PF_6^-$ is easily reduced to the Mo(0) complex by lithium aluminum hydride, equation (6-39).

$$[(\pi\text{-}C_5H_5)^-Mo^{++}(C_6H_5)]PF_6^- \xrightarrow{\text{LiAlH}_4} [(\pi\text{-}C_5H_5)^-Mo(C_6H_6)]Li^+ \quad (6\text{-}39)$$

Ring Expansion

An attempted Friedel–Crafts acylation of $(\pi\text{-}C_5H_5)(C_6H_6)M$ (M = Cr, Mn) resulted in expansion of the benzene ring to give the (π-tropylium)-(π-cyclopentadienyl) metal complexes, equation (6-40).

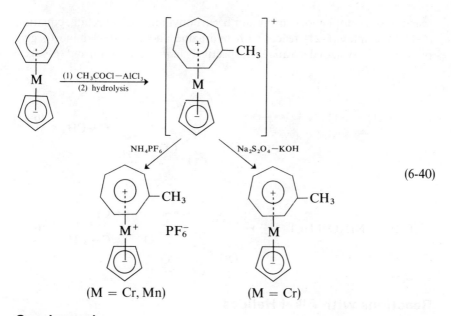

(6-40)

(M = Cr, Mn)   (M = Cr)

Condensation

A normal Claisen condensation of methylbenzoatechromium tricarbonyl and acetone occurs in a basic solution to give benzoylacetonechromium tricarbonyl, equation (6-41). This same complex can be obtained by the

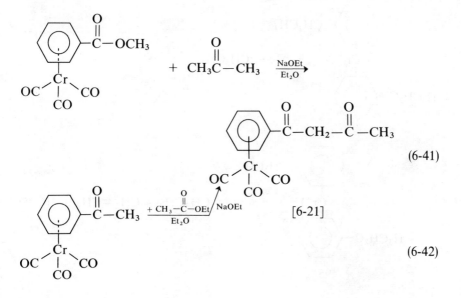

(6-41)

[6-21]

(6-42)

reaction of acetophenonechromium tricarbonyl and ethyl acetate, equation (6-42). Complex [6-21] reacts with hydrazine and hydroxylamine to give pyrazol and isoxazol derivatives, respectively, equations (6-43) and (6-44).

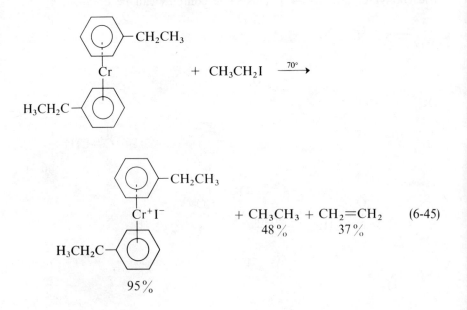

$$[6-21] + NH_2 \cdot NH_2 \longrightarrow \qquad (6\text{-}43)$$

$$[6-21] + NH_2OH \cdot HCl \longrightarrow \qquad (6\text{-}44)$$

Reactions with Alkyl Halides

Di-ethylbenzenechromium reacts with alkyl halides such as ethyl iodide and results in the oxidation of the central metal of the complex and the disproportionation of ethyl radicals derived from ethyl iodide, equation (6-45).

$$+ \; CH_3CH_2I \; \xrightarrow{70°}$$

$$+ \; CH_3CH_3 + CH_2{=}CH_2 \qquad (6\text{-}45)$$
$$ 48\% \qquad\quad 37\%$$

$$95\%$$

On the other hand, dimerization of the alkylradicals is observed with allyl (6-46), benzyl (6-47), and trityl halides. A possible mechanism for these reactions is proposed in equation (6-48).

$$2(EtPh)_2Cr^0 + 2CH_2{=}CH{-}CH_2Br \longrightarrow$$

$$2(EtPh)_2Cr^+Br^- + CH_2{=}CH{-}CH_2{-}CH_2{-}CH{=}CH_2$$

$$\text{(no propylene)} \qquad (6\text{-}46)$$

$$3(EtPh)_2Cr^0 + 4PhCH_2Cl \longrightarrow$$

$$2(EtPh)_2Cr^+Cl^- + CrCl_2 + 2EtPh + 2PhCH_2CH_2Ph \qquad (6\text{-}47)$$

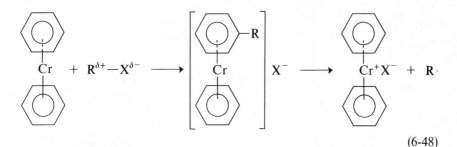

$$(6\text{-}48)$$

6-4. σ–π REARRANGEMENTS

σ–π Rearrangements involve the rearrangement of an organic group σ-bonded to a transition metal to the corresponding π-complex bonded species. σ-Bonding involves ligand-to-metal σ-orbitals (sp^n hybrid) on the ligand and d–$σ$-orbitals on the metal; π-complex bonding occurs by a π-overlap between the ligand π-orbitals and d–$π$-acceptor metal orbitals as well as a "back bond" resulting from a flow of electron density from filled metal d_{xy} or other $d_π$–$p_π$ hybrid orbitals into antibonding orbitals on the carbon atom (see Chapter 3). The π-complex bond thus involves a certain amount of double bond character in the metal–ligand interaction.

The first example of this rearrangement in organometallic chemistry came with the characterization of the π-arene chromium complexes, Hein's complexes, and the isolation of the intermediate in their preparation, tri-σ-phenylchromium tristetrahydrofuranate. Since then such rearrangements have been observed in organometallic compounds of almost all the transition metals and with a large variety of ligands, including groups bound through an element other than carbon. Possibly the most important application of σ–π rearrangements is in the field of homogeneous catalytic reactions such as

Ziegler–Natta polymerization of olefins and the oxo process—hydroformyl-ation of olefins (Chapter 7).

It is of value to examine closely some representative types of σ–π re-arrangements, subsequently applying some of the principles to known catalytic processes. Triphenylchromium tristetrahydrofuranate is a moder-ately stable purple crystalline complex. σ-Bonded Cr(III), d^3, is hexaco-ordinated to the three phenyl groups and three tetrahydrofuran ligands in an octahedral configuration [6-22], equation (6-49).

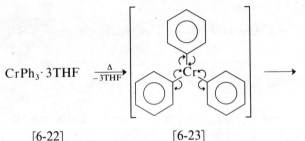

$$CrPh_3 \cdot 3THF \xrightarrow[-3THF]{\Delta}$$

[6-22]   [6-23]

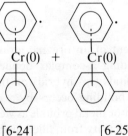

[6-24]   [6-25]   [6-26]

(6-49)

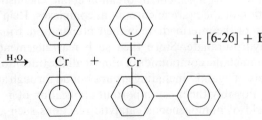

$\xrightarrow{H_2O}$ + [6-26] + Biphenyl

When this complex is heated the stabilizing THF ligands are easily removed. Homolytic cleavage of phenyl–chromium bonds in triphenyl chromium [6-23] results in the formation of three radical arene intermediates ([6-24], [6-25], [6-26]) and the reduction of Cr(III) to Cr(0). Hydrolysis gives π-dibenzene chromium(0), π-phenyl π-biphenyl chromium(0), π-bis-diphenyl chromium(0), and biphenyl. The significant point here is that Cr(III) was reduced to a zero valent d^6 metal that can satisfactorily accommodate 12 ligand electrons to give a stable d^2sp^3 octahedral metal π-complex. In summary, within the reaction media, an undistorted d^3, chromium(III), σ-bonded octahedral complex was rearranged to a more stable d^6, chromium-(0), π-bonded octahedral species.

This basic mechanism is supported whether we start with isolable σ-bonded triphenylchromium tritetrahydrofuranate or whether phenyl magnesium bromide and chromous chloride are reacted in tetrahydrofuran (one method of preparing the metal π-complexes).

However, if we react phenyl magnesium bromide in the presence of bromobenzene and a trace of cobaltous chloride (Kharash aryl coupling reaction), biphenyl is produced catalytically. This is quite unusual because alkyl or aryl halides do not react with arylmagnesium compounds or with metal halides, and metallic halides couple stoichiometrically—not catalytically—with aryl Grignard reagents. Here again the first step may involve the formation of a σ-bonded phenyl cobalt(II) complex. However, in this case we anticipate a d^7, generally labile, distorted tetrahedral cobalt complex. Thus, under the reaction conditions, the freshly reduced Co(0) behaving as a d^9 metal is not expected to form a stable d^2sp^3 octahedral diphenyl cobalt π-complex. Hence the active zero valent cobalt atom is free to react with bromobenzene, producing phenyl radicals and cobaltous bromide. The cobaltous bromide in turn may react with phenylmagnesium bromide, which is in excess, and the reaction thus proceeds in a catalytic fashion.

Similar mechanisms, based on the stability and characteristics of σ- vs π-complexes, have been prepared for catalytic, stoichiometric, and double-coupling reactions.

σ–π Rearrangements have been observed for σ-tricyclopentadienyliron and σ-dicyclopentadienyliron chloride, which rearrange to ferrocene and ferrocenium chloride, as given in equations (6-50) and (6-51).

We can anticipate that a ligand-associated (above $-60°$) Fe(III) complex represents a d^5 distorted octahedral, d^2sp^3, configuration that is labile and of very moderate stability. Thus a σ–π rearrangement giving an extremely stable ferrocene, Fe(II), d^6 undistorted octahedral d^2sp^3 configuration is easily visualized. Similarly, a stable Fe(III) d^5 ferricenium chloride also may be anticipated.

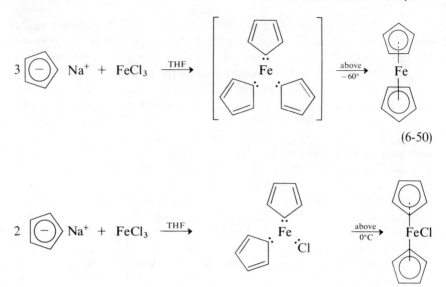

$$(6\text{-}50)$$

$$(6\text{-}51)$$

Analogous σ–π rearrangements are exemplified in reactions (6-52), (6-53), and (6-54).

$$(\pi\text{-}C_5H_5)Fe(CO)_2(\sigma\text{-}C_5H_5) \xrightarrow{\text{H}^+\text{(HCl)}} \left[\begin{array}{c} OC-Fe-CO \end{array} \right]^+ PF_6^- \qquad (6\text{-}52)$$

It should be noted that protonation of [6-27] produces [6-29]. Since protonation with deuterium chloride occurs at the 3-position, a mechanism involving a carbonium ion and its charge transfer to the metal is probably part of the reaction, equation (6-55).

However, the possibility of a hydride mechanism is not excluded. Irradiation of [6-27] may homolytically cleave the metal–carbon bond to form a π-allyl complex [6-28]. Hydride ion addition to the π-complex [6-29] gives an isopropyl complex, $C_5H_5Fe(CO)_3$—$CH(CH_3)_2$. This process is designated as a π–σ rearrangement. Analogous π-allyl and olefin complexes of molybdenum and of tungsten are known.

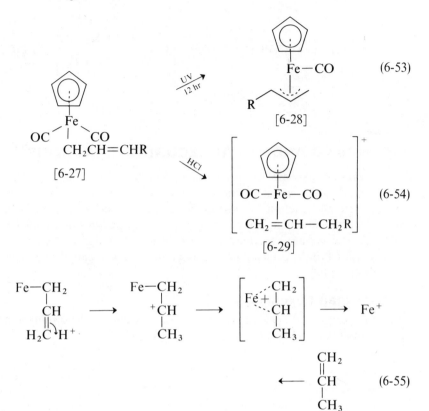

The reversible σ–π rearrangement given in equation (6-56) has been extensively studied:

$$C_5H_5Fe(CO)_2Et \underset{+H^-}{\overset{-H^-}{\rightleftharpoons}} \left[C_5H_5Fe-\underset{CO}{\overset{CO}{\underset{|}{\overset{|}{}}}} \overset{CH_2}{\underset{CH_2}{\|}} \right]^+ \quad (6\text{-}56)$$

[6-30]   [6-31]

If $Ph_3CClO_4^-$ acid is used, the rearrangement, equation (6-56), occurs; however, the use of Ph_3CCl causes ethyl group cleavage. The iron is oxidized, and the coupled product $[C_5H_5Fe(CO)_2]_2$ is formed.

The deuterated σ-complex [6-32] was reacted with Ph_3C^+ to give [6-33], equation (6-57). This result provides evidence that the hydride ion is eliminated from a β-carbon atom in this σ–π arrangement.

$$C_5H_5Fe(CO)_2CO(CH_3)_2 \xrightarrow{-H^-} \left[\begin{array}{c} CO \\ | \quad CH_2 \\ C_5H_5Fe-\| \\ | \quad C-CH_3 \\ CO \quad | \\ O \end{array} \right]^+ \quad (6\text{-}57)$$

[6-32]   [6-33]

6-5. LIGAND AND METAL EXCHANGE REACTIONS

Coordination complexes often undergo both single and multiple ligand exchanges or replacement reactions. Metal k-complexes also undergo analogous exchange reactions. For simplification we can consider arene, olefin, and metallocene exchange reactions separately. In this section we also consider metal exchange reactions. The implications of ligand exchange are closely related to the principles presented under metal π-complexes as intermediates in catalysis.

Arene Ligand Exchange

Both rings in dibenzenechromium can be exchanged by toluene and carbon monoxide under pressure, equation (6-58).

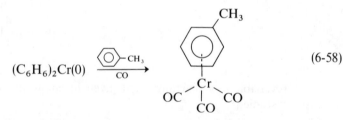

$$(C_6H_6)_2Cr(0) \xrightarrow[CO]{\langle\text{O}\rangle-CH_3} \qquad (6\text{-}58)$$

Dibenzenechromium is converted to dibiphenylchromium on similar treatment, and the reaction is irreversible under these conditions. The equilibrium is influenced by the concentration of the reactants and the relative stability of the products.

Benzene-biphenylchromium cation, however, undergoes a ligand exchange with other aromatic ligands (*in vacuo* or under nitrogen and in the presence of aluminum chloride) to yield the corresponding diarenechromium complexes, equation (6-59).

$$(C_6H_6)(C_6H_5{-}C_6H_5)Cr(I) \xrightarrow[\text{AlCl}_3]{\text{arene}} (\text{arene})_2Cr(I) \qquad (6\text{-}59)$$

Arene = biphenyl, benzene, mesitylene, tetralin,
and p,p'-dimethyl biphenyl

The exchange reaction of an arene ligand in $(C_6H_6)Cr(CO)_3$ in n-heptane is studied by using ^{14}C-labeled benzene $(^{14}C_6H_6)$. The rate of the exchange reaction is as shown in Fig. 6-7.

During the course of the formation of an intermediate [6-34] the slow and rate-determining step occurs. Structure [6-34] then decomposed to C_6H_6 and $Cr(CO)_3$. When two fragments combine to form $C_6H_6Cr(CO)_3$, the exchange reaction occurs.

When structure [6-35] is formed, the probability for the exchange reaction is increased.

The stability of arene carbonyls toward an exchange reaction, equation (6-60), is given by benzene < toluene < cycloheptatriene ≪ naphthalene.

$$Rate = k[C_6H_6Cr(CO)_3]^2 + k'[C_6H_6Cr(CO)_3][C_6H_6]$$

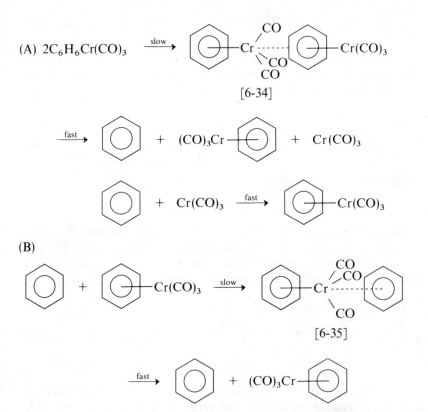

Fig. 6-7.

$$ArCr(CO)_3 + Ar^* \rightleftharpoons Ar^*Cr(CO)_3 + Ar \qquad (6\text{-}60)$$

$$Ar = arene$$

Olefin and Acetylene Ligand Exchange

Olefin exchange reactions are particularly related to catalytic reactions involving metal π-complexes. One of the first steps in the hydrogenation, isomerization, and polymerization of olefins involves the exchange of a ligand or solvent molecule to form the corresponding metal π-complex intermediate.

The exchange of an ethylene ligand in Zeise's salt was studied by NMR techniques. In contrast to the stability of the salt, the exchange was found to be very fast on the NMR time scale since only one sharp signal is observed when free ethylene and the salt are interacted, equation (6-61).

$$\left[\begin{array}{c} CH_2 \\ \| \text{--------} PtCl_3 \\ CH_2 \end{array} \right]^- + C_2^*H_4 \rightleftharpoons \left[\begin{array}{c} {}^*CH_2 \\ \| \text{--------} PtCl_3 \\ CH_2 \end{array} \right]^- + C_2H_4 \quad (6\text{-}61)$$

π-$t\cdot t\cdot t$-Cyclododecatrienenickel reacts with 3 moles of butadiene at $-40°C$, and the cyclododecatriene ligand is replaced by a newly formed linear butadiene-trimer, equation (6-62). When this reaction is carried out in

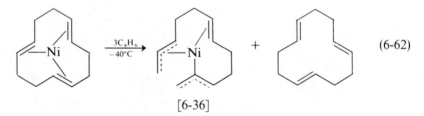

$$(6\text{-}62)$$

$$[6\text{-}36]$$

the presence of an excess of butadiene at 20°, it proceeds catalytically to give the $t\cdot t\cdot t$-cyclododecatriene through the intermediate complex [6-36]. The final step of the cyclization reaction may involve an internal electron shift process, equation (6-63), called "allyl coupling reaction."

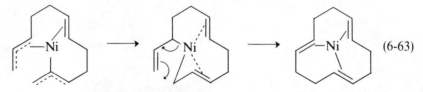

$$(6\text{-}63)$$

π-Cyclopentadienyl-π-1,5-cyclo-octadiene cobalt, $(\pi\text{-}C_5H_5)Co(C_8H_{12})$, undergoes an exchange reaction in the presence of 2 moles of tolane (diphenylacetylene). This reaction not only involves ligand exchange but also the cyclodimerization of tolane to tetraphenylcyclobutadiene, equation (6-64).

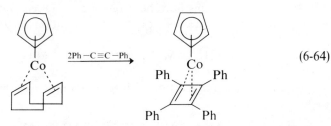

$$2Ph-C\equiv C-Ph \qquad (6-64)$$

Acetylenes often displace one another in metal π-complexes, equation (6-65). Electron-withdrawing groups impart stability to these complexes. The effectiveness increases in the order of acetylene < alkylacetylene < aryl-acetylene < p-nitroaryl-acetylene.

$$[Pt(acetylene\ A)(PPh_3)_2] + acetylene\ B \longrightarrow$$

$$[Pt(acetylene\ B)(PPh_3)_2] + acetylene\ A \qquad (6-65)$$

π-Cyclopentadienyl Exchange Reactions

Several exchange reactions involve the transfer of a π-cyclopentadienyl group from iron to palladium, nickel, cobalt, and titanium, the last exchange being reversible. The reported yields were quite high in some cases. A typical reaction is given in equation (6-66).

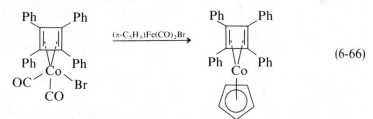

$$(\pi\text{-}C_5H_5)Fe(CO)_2Br \qquad (6-66)$$

The treatment of ferrocene with aluminum powder, aluminum chloride, and benzene results in the exchange of benzene for the cyclopentadienyl ligand to give a [cyclopentadienyl benzene iron]$^+$ cation. Further ligand exchange is very difficult.

$$ML_1 + L_1^* \rightleftarrows ML_1^* + L_1 \qquad (6-67)$$

$$ML_1 + L_2 \rightleftarrows ML_2 + L_1 \qquad (6-68)$$

M = metal, L = ligand, and L* = isotope labeled ligand

Ligand exchange reactions can be summarized by the use of the general equations (6-67) and (6-68).

Analogous reactions have been observed between two different metal π-complexes. Such reactions are expressed in equation (6-69).

$$ML_1 + M'L_2 \rightleftharpoons ML_2 + M'L_1 \qquad (6\text{-}69)$$

In many cases, the reactions follow a modified pattern, as in equation (6-70).

$$ML_1 + M'L_2 \rightleftharpoons L_1ML_2 + M' \qquad (6\text{-}70)$$

Metal Exchange Reactions in Metal π-Complexes

Excellent examples of metal exchange reactions are provided by complexes of tetraphenylcyclobutadiene. When either iron pentacarbonyl or nickel carbonyl is reacted with tetraphenylcyclobutadiene palladium bromide, the corresponding metal is exchanged with the release of palladium(0) metal. The reaction proceeds only in aromatic solvents and seems to be generally applicable to a variety of transition metal carbonyls. As illustrated in Fig. 6-8, the first step is proposed to involve the formation of uncomplexed tetraphenylcyclobutadiene, a mixed metal halide carbonyl, carbon monoxide,

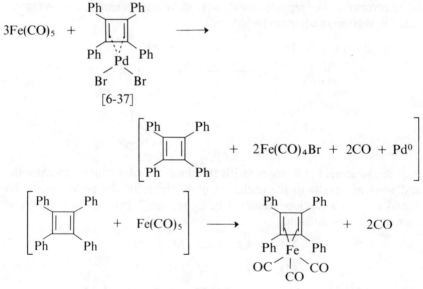

Fig. 6-8.

and palladium metal. Simultaneously, a second mole of metal carbonyl complexes with the olefin, giving the stable product. Incidentally, this final step represents a ligand exchange reaction.

Complex [6-37] also undergoes a similar exchange reaction with nickel carbonyl and cobaltocene, equations (6-71) and (6-72). This type of exchange reaction can be given by the general equation (6-73).

$$[6\text{-}37] + Ni(CO)_4 \longrightarrow \quad\text{(structure)}\quad NiBr_2 \qquad (6\text{-}71)$$

with Ph groups on the cyclobutadiene ring.

$$[6\text{-}37] + Co(C_5H_5)_2 \longrightarrow \quad\text{(structure)}\quad Co \qquad (6\text{-}72)$$

$$[6\text{-}37] + M(CO)_n \longrightarrow \quad\text{(structure)}\quad M(CO)_{n-2} \qquad (6\text{-}73)$$

Reactions (6-74) and (6-75) also are known.

$$\text{(structure)}\ MBr_2 + C_5H_5Fe(CO)_2Br \xrightarrow{\ \text{boiling } C_6H_6\ } \text{(structure)}\ M \qquad (6\text{-}74)$$

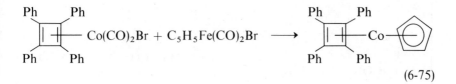

$$\text{(structure)}\ Co(CO)_2Br + C_5H_5Fe(CO)_2Br \longrightarrow \text{(structure)}\ Co \qquad (6\text{-}75)$$

Complex [6-38] is rather stable, and it also can be prepared by exchange with the anologous palladium complex, equation (6-76).

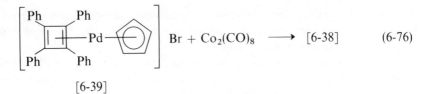

$$\left[\begin{array}{c} Ph \quad Ph \\ \text{[structure]} \text{—Pd—} \\ Ph \quad Ph \end{array} \right] Br + Co_2(CO)_8 \longrightarrow [6\text{-}38] \qquad (6\text{-}76)$$

[6-39]

6-6. VALENCE TAUTOMERISM AND VALENCE ISOMERISM

In solution some olefin π-complexes have been found to exist in dynamic forms. Cyclo-octatetraeneiron tricarbonyl, $(C_8H_8)Fe(CO)_3$, serves as an example. The $Fe(CO)_3$ group in this complex shifts about the cyclo-octatetraene ring (Fig. 6-9).

These structures, which possess different valence bonds, are isomers. The phenomenon is called valence tautomerism (also referred to as ring whizzing). Other species also are known to exhibit valence tautomerism in solution (Fig. 6-10).

If the energy barrier between two forms is small, we may anticipate valence tautomerism. However, if the energy barrier is high, we are dealing with valence isomerism.

At room temperature only one MNR peak was observed for π-C_7H_7V-$(CO)_3$. This was interpreted to mean that the entire seven-electron system complexed with the vanadium metal atom [6-41] (delocalized molecular

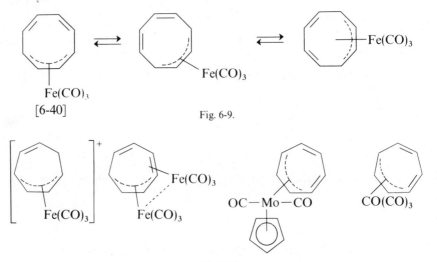

Fig. 6-9.

Fig. 6-10.

orbital). However, at 160°C the NMR spectra reveals a complex broad band, indicating a structure such as [6-42] in which the three triple bonds are localized. The degree of symmetry has been decreased as the type of bonding was altered.

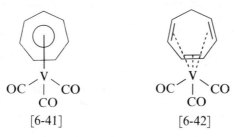

[6-41]   [6-42]

A final example of valence isomerism involves a σ–π rearrangement. π-Allyl palladium chloride [6-43] becomes a σ-allyl complex in dimethyl sulfoxide as determined by NMR study. The rearrangement is facilitated by the interaction of the solvent with the metal. When a stronger donor ligand, such as PPh$_3$, was used in different ratios (1:1 and 2:1), structures [6-44] and [6-45] were obtained, equations (6-77 and 6-78). A σ-allylic complex [6-45] exists with [6-46] as a valence tautomer.

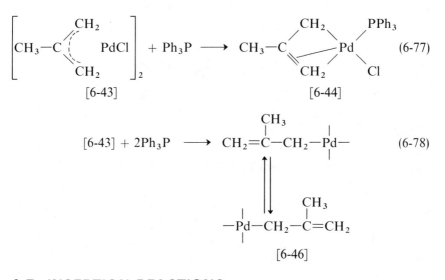

6-7. INSERTION REACTIONS

Insertion reactions have great potential in organic synthesis. This section briefly outlines some of the more useful insertion reactions.

Carbon Monoxide Insertion

σ-Allyl metals, for example, $R-Mn(CO)_5$ and $RCo(CO)_4$, readily undergo insertion of CO into the $R-M$ bond by reaction with CO or with a donor ligand such as R_3P, equations (6-79) and (6-80).

$$R-Mn(CO)_5 + CO \longrightarrow R-\overset{\parallel}{\underset{O}{C}}-Mn(CO)_5 \qquad (6\text{-}79)$$

$$R-Mn(CO)_5 + R_3P \longrightarrow R-\overset{\parallel}{\underset{O}{C}}-Mn(CO)_4PR_3 \qquad (6\text{-}80)$$

π-Allyl complexes also may undergo insertion reactions (6-81), that is, when coupling of two allyl groups does not occur (6-82).

$$\left[\left\langle\!\!\!\left\langle \; PdCl \right.\right]_2 + CO \xrightarrow{\;HCl\;} \overset{\displaystyle \left\langle\!\!\!\left\langle \right.}{CH_2COCl} \qquad (6\text{-}81)$$

$$\left| \left\langle\!\!\!\left\langle \; \right.\right|_2 Ni + CO \longrightarrow biallyl \qquad (6\text{-}82)$$

Carbon monoxide usually displaces the olefin in olefin π-complexes. A similar displacement reaction occurs with π-C_5H_5 and π-C_6H_6 complexes.

Acetylene Insertion

π-Allyl complexes sometimes undergo acetylene insertion reactions, as given in equation (6-83).

$$\left[\left\langle\!\!\!\left\langle \; NiCl \right.\right]_2 + C_2H_2 + CO \xrightarrow{\;ROH\;} \diagdown\!\!\diagup\!\!\diagdown\!\!\diagup^{CO_2R} \qquad (6\text{-}83)$$

$$\left\langle\!\!\!\left\langle -M + R-C\equiv C-R \longrightarrow \underset{M}{\overset{R-C\equiv C-R}{\diagup}} \longrightarrow \right.$$

$$\underset{R\quad M}{\overset{R}{\diagdown\!\!\diagup\!\!\diagdown}} \qquad (6\text{-}84)$$

However, the insertion probably proceeds through a σ-allyl derivative since π–σ conversion is anticipated upon coordination of an acetylene group to the metal, equation (6-84).

Olefin Insertion

Olefin insertion to π-allyl derivatives similarly may proceed through a σ-allyl intermediate. A good example is given by the formation of 1,4-hexadiene from ethylene and butadiene in the presence of rhodium chloride, equation (6-85). (See Zeigler–Natta catalyst in Chapter 7).

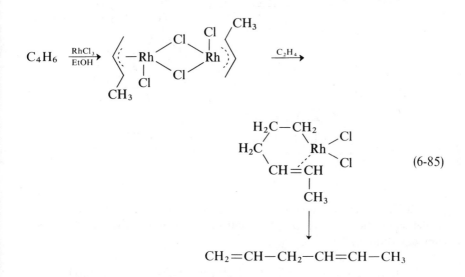

$$CH_2=CH-CH_2-CH=CH-CH_3$$

(6-85)

6-8. IRRADIATION OF METAL π-COMPLEXES

One of the most recently studied reactions are those stimulated by γ-irradiation. Irradiation techniques also find application in the preparation of novel as well as some more common metal π-complexes. The γ-irradiation induced reactions of ferrocene in a variety of halogenated solvents have been studied. As shown in equations (6-86) and (6-87), ferrocene irradiated in the presence of carbon tetrachloride or carbon tetrabromide gives ferricinium tetrachloroferrate(III) and ferriciniumtetrabromoferrate(III), respectively.

When π-dicyclopentadienyl titanium dichloride is irradiated in the presence of carbon tetrachloride, π-cyclopentadienyl titanium trichloride is obtained in sufficient yield to be considered a new preparative method, equation (6-88).

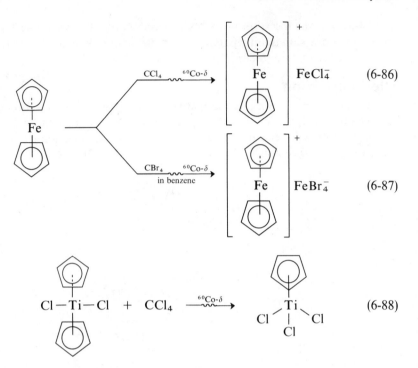

$$(6\text{-}86)$$

$$(6\text{-}87)$$

$$(6\text{-}88)$$

Dibiphenylchromium(I) cation disproportionates slowly upon ultraviolet irradiation, giving dibiphenylchromium(0) and a divalent chromium ion, equation (6-89).

$$(C_6H_5-C_6H_5)_2Cr^+ \xrightarrow[(+\text{dipy})]{h\nu} (C_6H_5-C_6H_5)_2Cr^0 + [Cr(\text{dipy})_3]^{2+}$$

$$(6\text{-}89)$$

A novel reaction was observed in the conversion of one metal-arene π-complex to another by thermal neutron bombardment. In this way dibenzenemolybdenum was converted to dibenzenetechnetium(I), equation (6-90).

$$Mo(C_6H_6)_2 \xrightarrow{n\delta} {}^{99}Mo(C_6H_6)_2 \xrightarrow{-\beta} {}^{99}Tc(C_6H_6)_2^+ \qquad (6\text{-}90)$$

6-9. NOVEL APPLICATIONS OF METAL π-COMPLEX REACTIONS

Metal π-complex formation is utilized for such seemingly unrelated applications as the separation or purification for analysis, resolution of racemic mixtures, and the stabilization of unstable intermediates.

Separation and Purification of Olefins and Acetylenes

The stability constants of a great number of mono- and polyolefin complexes of silver nitrate have been determined by gas chromatographic studies. The results of these studies are summarized in Table 6-4. These observations can be utilized for isolation and purification. For example, *trans*-cyclo-octene is separated from the *cis*-isomer by the preferential formation of the more stable *trans*-cyclo-octene silver complex. The very strained olefin, cyclo-propene, is quantitatively removed from a mixture of olefins by passing it with nitrogen through an aqueous silver nitrate solution.

Table 6-4. Order of Stability of Olefin
π-Complexes

Terminal olefin	$\longrightarrow$	internal olefin
cis-Olefin	$\longrightarrow$	*trans*-olefin
More-strained olefin	$\longrightarrow$	less-strained olefin

The silver salts of acetylene, which are suspected to be silver complexes (Fig. 6-11), are usually insoluble and are utilized in purification or separation.

The acetylene complex of zero valent platinum $(R-C\equiv C-R)Pt^0(PPh_3)_2$ is a useful reagent for the extraction of acetylenic compounds from natural products.

Olefins often form σ-bonded organomercury compounds via a mercury π-complex intermediate. Since this reaction proceeds quite readily and quantitatively, it may be used as an analytical method for the detection of olefins.

Resolution of Racemic Compounds

π-Complex formation was ingeniously utilized for the separation of racemic *trans*-cyclo-octene. Since *trans*-cyclo-octene has a rigid conformation because of the presence of a *trans*-double bond in this middle-sized ring, it was felt that resolution of its optical isomers was possible. The displacement of an ethylene ligand in the Zeise salt by *trans*-cyclo-octene proceeds smoothly because the strained *trans*-olefin forms a more stable complex and the expelled ethylene is eliminated from the reaction media as a gas, equation (6-91).

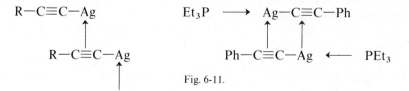

Fig. 6-11.

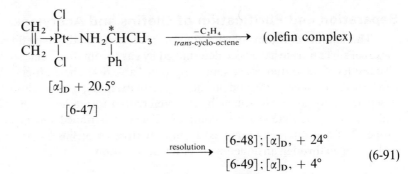

$$[\alpha]_D + 20.5°$$

$$[6\text{-}47]$$

$$\xrightarrow{\text{resolution}} \quad \begin{array}{l} [6\text{-}48]; [\alpha]_D, + 24° \\ [6\text{-}49]; [\alpha]_D, + 4° \end{array} \qquad (6\text{-}91)$$

The resulting complex can be separated into diastereoisomers [6-48] and [6-49]. *Trans*-cyclo-octene was liberated from the separated complexes by the addition of an aqueous potassium cyanide solution, equations (6-92) and

$$[6\text{-}48] \xrightarrow{\text{aq. KCN}} \textit{trans}\text{-cyclo-octene } [\alpha]_D - 21° \qquad (6\text{-}92)$$

$$[6\text{-}49] \xrightarrow{\text{aq. KCN}} \textit{trans}\text{-cyclo-octene } [\alpha]_D + 18.5° \qquad (6\text{-}93)$$

(6-93). In this way, the presence of optical isomers of *trans*-cyclo-octene was demonstrated. A so-called "*trans, trans*-1,5-cyclo-octadiene" was resolved in an analogous manner into optical isomers $D^{-26°}$ and $D^{+34°}$, respectively. However, *trans, trans*-1,5-cyclo-octadiene has a plane of symmetry. Therefore, the compound could be identified as *cis, trans*-1,5-cyclo-octadiene, which has no plane of symmetry and is asymmetric.

Stabilization of Unstable Nonbenzenoids

Probably the most academically dramatic use of metal π-complex formation was demonstrated by the trapping of the cyclobutadiene–silver nitrite complex. Many subsequent cyclobutadiene derivatives were isolated and proved to be rather stable complexes. For example, tetraphenylcyclobutadieneiron tricarbonyl melts without decomposition at 234° and tetraphenylcyclobutadiene π-cyclopentadienyl cobalt melts at 256°C under nitrogen. Upon examination of the electronic configuration the stability of these complexes often can be predicted.

Unsubstituted cyclobutadiene and benzocyclobutadiene also were trapped as the iron tricarbonyl π-complexes.

Attempts to liberate stable cyclobutadiene from its π-complexes have been unsuccessful. However, good evidence have been obtained for the existence of free cyclobutadiene at liquid nitrogen temperature. Free cyclo-butadiene liberated from the oxidation of cyclobutadieneiron tricarbonyl reacts with many olefins and acetylenes to give addition products.

Stabilization of cumulenes by π-complex formation with an iron tricar-bonyl group serve as further examples:

$$H_2C=C=C=CH_2 \cdot Fe_2(CO)_6$$

$$CH_3=CH=C=C=CH-CH_3 \cdot Fe_2(CO)_6$$

$$(CH_3)_2C=C=C=C(CH_3)_2 \cdot Fe(CO)_6$$

Unsubstituted butatriene is unstable at 0°, but the complex decomposes only above 230° under nitrogen. Allyl and tropylium ions were also trapped as metal complexes ([6-50], and [6-51]). O-Quinodimethene iron carbonyl com-plex also serves as the metal-stabilized active intermediate [6-52].

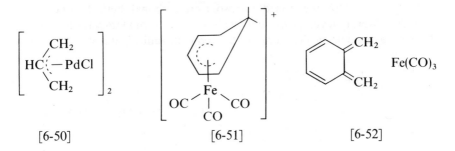

[6-50]   [6-51]   [6-52]

There have been many attempts to trap benzyne intermediates. The iso-lation of the cyclic acetylene π-complex, perfluorocyclohexyne-1-ene-3 [6-53],

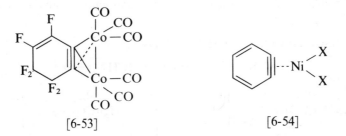

[6-53]   [6-54]

stimulated further interest in benzyne isolation. Subsequently, two types of nickel benzyne complexes have been isolated: [6-54] and [6-55].

π-Cyclopentadienyl hexakistrifluoromethyl benzene rhodium [6-56], a nonplanar benzene moiety, has been stabilized as a metal π-complex. Only four carbon atoms of the benzene ring are involved in bonding to the metal

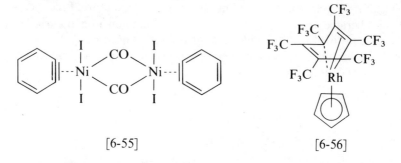

[6-55]   [6-56]

ion. C_1 and C_4 form σ-bonds, while the π-electrons between C_2 and C_3 form a π-complex bond with the metal. Nonplanarity of the benzene moiety may result from the steric interactions between adjacent CF_3 groups.

As an example of small ring stabilization, consider the reaction of either cobalt carbonyl or the mixed nitroso iron carbonyl $Fe(CO)_3NO^-$ with triphenyl-3-bromocyclopropene in acetonitrile at room temperature, [6-57]. Either the metal or the ligand can be altered with equivalent results.

$$\text{M = Co:}\quad x = y = CO$$
$$\text{M = Fe:}\quad x = CO; y = NO$$

[6-57]

Protonation of an olefin gives a carbonium ion that can be stabilized through metal π-complex formation. In reaction (6-94), the final π-cyclopentadienyl π-propylene iron dicarbonyl cation represents a stable d^6Fe octahedral d^2sp^3 complex.

$$(\pi\text{-}C_5H_5)(CO)_2Fe\text{—}CH_2CH=CH_2 \xrightarrow{\ H^+\ }$$

$$(\pi\text{-}C_5H_5)(CO)_2Fe\text{—}CH_2^+CHCH_3 \longrightarrow$$

(6-94)

$$\left[(\pi\text{-}C_5H_5)(CO)_2Fe\overset{+}{:}\begin{array}{c}\cdots CH_2\\ \big|\\ CH\\ \big|\\ CH_3\end{array}\right] \longrightarrow \left[(\pi\text{-}C_5H_5)(CO)_2Fe\leftarrow\begin{array}{c}CH_2\\ \|\\ CH\\ \big|\\ CH_3\end{array}\right]^+$$

Molecular Nitrogen Fixation

Molecular nitrogen fixation by microorganisms has been one of the mysteries in science. Although enzymes, the biological catalysts promoting

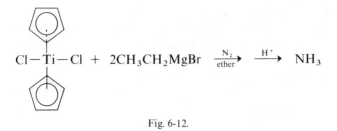

Fig. 6-12.

fixation, have been isolated, the role of the enzyme in nitrogen fixation has not been clarified. It has been said that some trace metal derivatives are required for the fixation. For years investigators have endeavored to understand this efficient biological catalysis in their attempts to fix nitrogen by chemical reactions.

Nitrogen fixation under milder conditions using a transition metal catalyst has been reported. The reaction is a simple one: Nitrogen was bubbled into a mixture of transition metal halides and ethyl magnesium halide in ether. A typical reaction is shown in Fig. 6-12.

A plausible mechanism has been postulated that includes the formation of titanocene dihydride intermediate as identified by ESR, Fig. 6-13.

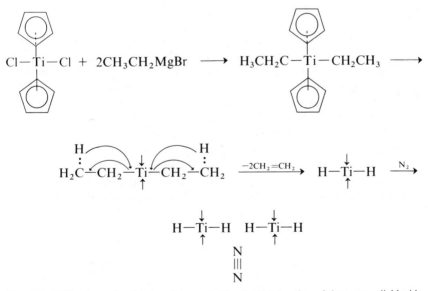

Fig. 6-13. Mechanism of molecular nitrogen fixation by the reaction of titanocene dichloride ethyl Grignard reagent and nitrogen.

   Diethyl titanocene is evidently being formed during the first step of the reaction. In an anologous fashion to the cleavage of the carbon–metal bond in organo-transition metal compounds, homolytic cleavage between titanium metal and ethyl carbon atoms gives titanium hydride and ethylene (β-hydrogen atoms are usually labile).

6-10. EXERCISES

6-1.   Discuss the effects of organic substituents in ferrocene on the entering group.

6-2.   Suggest a possible mechanism for the formulation of $(\pi\text{-}ArC_5H_4)Fe(\pi\text{-}C_5H_5)$, from the reaction of ferrocene and an aryldiazonium salt.

6-3.   Discuss the bonding in ferrocene based on its reactivity.

6-4.   Suggest methods for the preparation of the following compounds starting with ferrocene or cyclobutadieneiron tricarbonyl.

   a. $(\pi\text{-}C_5H_5)(\pi\text{-}C_5H_4COR)Fe$

   b. $(\pi\text{-}C_5H_5)(\pi\text{-}C_5H_4R)Fe$

   c. $(\pi\text{-}C_5H_5)(\pi\text{-}C_5H_4SO_3H)Fe$

   d. $(\pi\text{-}C_5H_5)(\pi\text{-}C_5H_4CH_2OH)Fe$

   e. $(\pi\text{-}C_4H_3O)Fe(CO)_3$

6-5.   What important factor is responsible for the fact that metallocenes undergo aromatic substitution reactions while arene complexes do not?

6-6.   Although arene π-complexes such as dibenzene chromium do not undergo electrophilic substitution reactions, benzenechromium tricarbonyl does undergo Friedel–Crafts acylation. Suggest a possible reason for this fact.

6-7.   What evidence is there for radical formation (homolytic cleavage) in the reaction $(\sigma\text{-}aryl)_xM \rightarrow (\pi\text{-}arene)_xM$?

6-8.   Reversible σ–π rearrangements are common. Suggest conditions and give an example of this reversible-type reaction.

6-9.   Ligand exchange reactions involving arenes, olefins, and acetylenes occur. Give examples of all their exchange reactions.

6-10. What is structurally different about ferrocene and ruthenocene?

6-11. Show a method for the preparation of the following:

   a. Homoanular-bridged ferrocene

   b. Heteroanular-bridged ferrocene

6-12. Why does anhydrous acetic acid react approximately 4.5 times as fast with vinylosmocene as with vinylferrocene?

6-13. Metallocenes can be reduced by various reagents. Complete the following reactions:

    a. $(\pi\text{-}C_5H_5)_2Co + NaBH_4 \longrightarrow$

    b. $(\pi\text{-}C_5H_5)_2Co^+ + NaBH_4 \longrightarrow$

    c. $(\pi\text{-}C_5H_5)(CO)_3Fe^+ + NaBH_4 \longrightarrow$

    d. $(\pi\text{-}C_5H_5)_2Ni + NaHg \xrightarrow{EtOH}$

6-14. Complete the following reactions:

    a. $(\pi\text{-}6_6H_6)Re(II) + 2PhLi$

    b. $(\pi\text{-}C_6H_6)Mn(CO)_3^+ + PhLi$

6-15. Arene complexes undergo condensation reactions. Predict the products
from the 2 following sets of reactants.

    a. $(\pi\text{-}C_5H_5COCH_3)Cr(CO)_3 + CH_3COOEt \xrightarrow[Et_2O]{NaOEt}$

    b. $(\pi\text{-}C_5H_5COCH_3)Cr(CO)_3 + CH_3COCH_3 \xrightarrow[Et_2O]{NaOEt}$

6-16. Predict the products of the following reactions:

    a. $(\pi\text{-}Ph_4C_4)PdBr_2 + M(CO)_x \longrightarrow$

    b. $(\pi\text{-}Ph_4C_4)PdBr_2 + (\pi\text{-}C_5H_5)_2Co \longrightarrow$

    c. $(\pi\text{-}Ph_4C_4)MBr_2 + (\pi\text{-}C_5H_5)Fe(CO)_2Br \xrightarrow{boiling\ C_6H_6}$

    d. $R\text{-}Mn(CO)_5 + CO \longrightarrow$

    e. $R\text{-}Mn(CO)_5 + Ph_3P \longrightarrow$

    f. $[(\pi\text{-}C_3H_5)NiCl]_2 + C_2H_2 + CO \xrightarrow{ROH}$

    g. $(\pi\text{-}C_5H_5)_2TiCl_2 + CCl_4 \xrightarrow{^{60}Co\text{-}\delta\ irradiation}$

6-17. The ability of olefins and acetylenes to form π-complexes can be used
for several important processes. Give examples of these phenomena.

6-18. Direct chlorination and nitration on ferrocene have not been reported.
Why?

6-19. The following reaction is reversible:

$$\text{Ferrocene} \rightleftarrows \text{ferracinium cation}$$

Suggest the reagents for these reactions.

6-20. Prepare the following compounds from ferrocene (Fc).

    a. $Fc\text{-}C{\equiv}C\text{-}Fc$

    b.

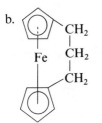

c.

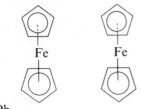

d. (Fc)$_4$Pb

6-21. Nickellocene dimerizes ethylene to 1-butene under 500 psi at 200°C; however, ferrocene does not. Why?

6-22. Describe a method for the preparation of the following metallocene hydrides:

a.   b.

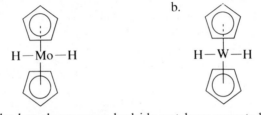

Why has chromocene hydride not been reported?

6-23. Describe the preparation of the following complexes:

a. $CH_2{=}CHFe(CO)_2C_5H_5$

b. $C_6F_5Fe(CO)_2C_5H_5$

c.

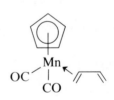

d.

$OC-Mn$

6-24. Complete the following reactions:

a. $[Pt(CH_2{=}CH_2)Cl_2]_2 \xrightarrow{CO}$

b. $[(HC{=}CHR)PdCl_2]_2 \xrightarrow{CO}$

6-25. Synthesize the following compounds from cyclobutadiene iron tricarbonyl:

a. Dewar benzene

b. Cubane

c. Cyclo-octatriene

6-26. Dibenzenechromium iodide is very soluble in water. Benzene-biphenyl-chromium iodide is fairly soluble in water, but bis-biphenylchromium iodide is insoluble. Why?

6-27. How does aluminum chloride react with dibenzenechromium(0) and ferrocene?

6-28. Dibenzenechromium(0) is easily oxidized in air to its cation. The cation can be reduced to the zero valent complex. Suggest several reducing agents.

6-29. What products can be expected from the following reaction:

Benzene + $PdCl_2$ $\xrightarrow{O_2}$

6-30. $\sigma-\pi$ Rearrangements involve cleavage of the bond between carbon and metal atoms. Discuss the nature of the bond cleavage.

6-31. Propose a plausible mechanism for the following reactions:

a. $3C_5H_5MgBr + FeCl_3 \longrightarrow$ ferrocene

b. $2C_5H_5MgBr + FeBr_2 \longrightarrow$ ferrocene

6-32. Completion of the following reaction takes about 6 months because the reaction temperature cannot be raised. Why?

$$2C_5H_5Na + FeCl_3 \xrightarrow[-80°]{THF} (\sigma\text{-}C_5H_5)_2FeCl\cdot THF$$

6-33. Write all products from the following reactions:

$$3 \langle\bigcirc\rangle\text{---}MgBr + CrCl_3 \xrightarrow[CO]{THF}$$

6-34. Complete the following reactions:

a. Dibenzenechromium(0) $\xrightarrow[\text{pressure (under } N_2)]{CO}$

b. Benzenechromium tricarbonyl $\xrightarrow{\text{benzene}}$

c. Ferrocene + benzene $\xrightarrow{\Delta}$

d. Ferrocene + dibenzenechromium(0) $\xrightarrow[N_2]{\Delta}$

e. Dibenzenechromium chloride + bi-phenyl $\xrightarrow{AlCl_3}$

6-35. Are the following reverse $\sigma-\pi$ rearrangements possible? State reasons for your prediction.

a. Ferrocene $\longrightarrow$ di-σ-cyclopentadienyl iron(II)

b. Dibenzenechromium(0) $\longrightarrow$ di-σ-phenylchromium(II)

6-36. NMR and IR spectroscopy are essential tools for the investigation of valence tautomerism. Predict the changes in the spectra of the following at 60°, room temperature, and $-60°$.

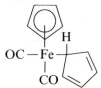

6-37. Dimethyl-*p*-phthalate can be prepared from the reaction of benzene and carbon monoxide in methanol in the presence of transition elements. Metal carbonyls also are excellent catalysts for this reaction. Design a mechanism for this reaction.

6-38. Methacrylate is catalytically prepared from acetylene, carbon monoxide, and methanol using a nickel carbonyl catalyst. Predict a plausible mechanism for this reaction.

6-39. Tris-triphenyl phosphine cobalt nitrogen complex, N_2Co $(PPh_3)_3$, reacts with ethylene to form an ethylene complex. Suggest a structure for this complex.

6-40. Show possible methods for trapping benzyne.

6-11. BIBLIOGRAPHY

C. W. Bird, "Transition Metal Intermediates in Organic Synthesis," Logos Press, London (1967).

G. E. Coates, M. L. H. Green, and K. Wade, "Organometallic Compounds," Vol. II, Methuen & Co., London (1968).

E. O. Fischer and H. P. Fritz, "Compounds of Aromatic Ring Systems and Metals," *in* "Advances in Inorganic Chemistry and Radio Chemistry," H. J. Emeleus and A. G. Sharpe (eds.), Vol. 1, Academic Press, New York (1959).

E. O. Fischer and H. Werner, "Metal π-Complexes," Vol. 1, Elsevier Publishing, Amsterdam (1966).

P. M. Maiths, "Cyclobutadiene–Metal Complexes," *in* "Advances in Organometallic Chemistry" F. G. A. Stone and Robert West (eds.), Vol. 4, Academic Press, New York (1966).

P. L. Pauson, "Compounds Derived from Cyclopentadiene," *in* "Non-Benzenoid Aromatic-Compounds" D. Ginsburg (ed.), Interscience Publishers, New York (1959).

P. L. Pauson, "Organometallic Chemistry," Edward Arnold, London (1967).

M. D. Rausch, *Cand. J. Chem.* **41** : 1289 (1963).

M. Rosenblum, "Chemistry of the Iron Group Metallocenes," Part One, Interscience Publishers, New York (1965).

G. Wilkinson and F. A. Cotton, "Cyclopentadienyl and Arene Metal Compounds," *in* "Progress in Inorganic Chemistry," F. A. Cotton (ed.), Vol. 1, Interscience Publishers, New York (1959).

H. Zeiss (ed.), "Organometallic-Chemistry," Reinhold Publishing, New York (1960).

H. Zeiss, P. J. Wheatley, and H. J. S. Winkler, Benzenoid–Metal Complexes, Ronald Press, New York (1966).

CHAPTER 7

Catalysis Involving Metal π-Complex Intermediates

In recent years it has become evident that many metal-catalyzed reactions proceed via a substrate metal π-complex intermediate. Commercially, the most significant of these include the polymerization of ethylene, the hydroformylation of olefins yielding aldehydes (oxo process), and the air oxidation of ethylene-producing acetaldehyde (Wacker process).

There are two major types of catalytic reactions: homogeneous and heterogeneous catalysis. The former requires that the substrate and catalyst react in a single phase, for example, liquid–liquid or gas–gas. The latter requires that at least a two-phase system is operative, for example, solid–liquid and liquid–gas. The distinction between homogeneous and heterogeneous processes is not always evident. Subtle differences or changes in solubilities, the often unclear role of intermediates, and the physical characteristics of the real catalytic species are not always apparent without some detailed examination.

Mechanisms involving metal π-complex intermediates have been postulated for both heterogenous and homogeneous catalytic systems. In heterogenous catalyses, π-complex formation between the catalyst surface and π-ligands have been spectroscopically studied. It has been proven that chemisorption between the catalyst and π-ligands initially occurs in most cases. Mechanisms involving the formation of π-complex intermediates in homogeneous catalysis have been studied in more detail than those involving heterogeneous catalysis. Most of the mechanisms postulated that include substitution reactions and *cis* ligand orientation effects have been adapted from the basic principles of coordination chemistry. Typical organometallic reactions such as σ–π and π–σ rearrangements also play important roles in homogeneous catalysis. It is interesting to note that some of the mechanisms postulated for homogeneous catalysis are applicable in some heterogeneous systems.

Throughout this chapter examples of mechanisms proposed for various catalytic processes that involve metal π-complex intermediates will be given. At the same time these examples will be utilized to present some of the basic principles of coordination chemistry. The first example involves the examination of the homogeneous hydrogenation of ethylene by tris (triphenylphosphine) rhodium(I) chloride, $RhCl(PPh_3)_3$ (Fig. 7-1). In this sequence there are a number of distinct processes. In going from structure [7-1] to [7-2] there is an illustration of metal oxidation of Rh(I) to Rh(III) and a square planar structure expanded to an octahedral dihydride. This process is called oxidative addition. The formation of [7-3] from [7-2] represents a substitution reaction and metal π-complex formation, as ethylene replaces the PPh_3 ligand. This is followed by the addition of a hydrogen atom to ethylene, the addition of a new ligand on Rh(III) and a π-to-σ rearrangement of the partially hydrogenated ethylene group (structures [7-3] and [7-4]). The final steps [7-4] to [7-1], probably consist of hydrogen addition to the σ-bound ethyl group followed by the ejection of ethane, ligand substitution, and reduction of the Rh(III) to Rh(I). These steps complete the hydrogenation process and are the essential features of the catalytic process. The first and last step—the oxidation of Rh(I) to Rh(III) and the reduction of Rh(III) to Rh(I) accompanied

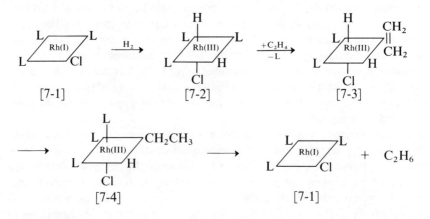

Fig. 7-1. Homogeneous hydrogenation of ethylene by $RhCl(PPh_3)_3$. L = PPh_3 as solvent.

by the conversion of the square planar structure to the octahedral structure—can be predicted on the basis of coordination chemistry.

Rh(I) has a d^8 outer shell electronic configuration and is likely to assume the stable dsp^2 square planar configuration since the large Rh(I) ligand field splitting usually causes a square planar coordination rather than a tetrahedral coordination. The d^6 Rh(III) ion assumes an octahedral configuration because of increased ligand field stabilization energy.

The actual hydrogenation [7-3] to [7-4] and [7-4] to [7-1] illustrates a *cis* ligand orientation effect. Two ligands oriented *cis* to each other in either a six or four coordinated species are considered favorably situated for subsequent reaction. In some cases a new ligand is formed as in [7-3] to [7-4], and in other cases a new compound is produced [7-4] to [7-1].

As indicated earlier, the other coordination principles involved shall be discussed in subsequent examples.

7-1. HYDROGENATION OF OLEFINS

Halpern has utilized the principles of reactivity or coordination compounds to explain reaction mechanisms, some examples of which involve metal π-complexes. His concept of hydrogenation of olefins is based upon the formation and cleavage of a hydrido transition metal complex.

Complexes of Cu(II), Cu(I), Ag(I), Hg(II), Hg(I), Co(I), Co(II), Pd(II), Pt(II), Rh(I), Rh(II), Ru(II), Ru(III), and Ir(I) have catalyzed homogeneous hydrogenation reactions in solution. In each case H_2 is split by the catalyst with the formation of a reactive transition-metal hydride (or hydrido) complex as an intermediate. Three distinct mechanisms have been advanced, as given below. The first mechanism involves heterolytic splitting:

(heterolytic-splitting)
$$[Ru(III)Cl_6]^{3-} + H_2 \;\rightleftharpoons\; [Ru(III)HCl_5]^{3-} + H + Cl^-$$

The assignment of oxidation numbers to the metal atoms above and in subsequent complexes follows the convention that the oxidation number (-1) is given to the hydride (hydrido) ligand. Heterolytic splitting involves a substitution process without change in the formal oxidation number of the metal. Thus reactivity is governed by the lability of the ligands toward substitution, the stability of the hydride formed, and the effectiveness of the base (which may be the solvent or the displaced ligand) in trapping the released proton. An example of heterolytic splitting is seen in the homogeneous hydrogenation of maleic acid with an aqueous solution of ruthenium dichloride. The heterolytic splitting of H_2 by a Ru(II) olefin complex is the rate-determining step (Fig. 7-2).

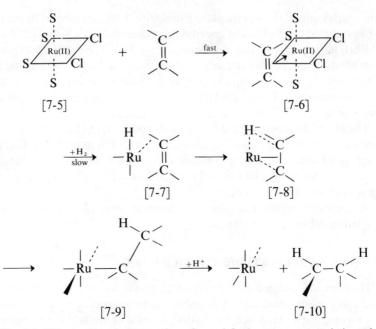

Fig. 7-2. Homogeneous hydrogenation of an olefin in an aqueous solution of ruthenium [7-5] octahedral structure derived from a square planar Ru(II) + 2s is an assumption) S = solvent.

The second and third mechanisms involve either homolytic splitting or oxidative addition:

(homolytic splitting)
$$2[Co(II)(CN)_5]^{3-} + H_2 \rightleftharpoons 2[Co(III)H(CN)_5]^{3-}$$

(oxidative addition—formation of a dihydride)
$$Ir(I)Cl(CO)(PPh_3)_2 + H_2 \rightleftharpoons Ir(III)H_2Cl(CO)(PPh_3)_2$$

In both these cases the metal is formally oxidized upon hydride formation and the reactivity is influenced by the ease with which the metal is oxidized. If we minimize the effect of ligand variation the tendency is toward oxidation; thus the reactivity toward H_2 for square planar d^8 complexes is Os(0) > Ru(0) > Fe(0) > Ir(I) > Rh(I) > Co(I) > Pt(II) > Pd(II) $\ll$ Ni(II), Au(III).

Homogeneous catalytic hydrogenation of an olefin via homolytic splitting of hydrogen is illustrated by the $[Co(CN)_5]^{3-}$-catalyzed hydrogenation of butadiene in aqueous solution (Fig. 7-3).

Both ethylene and acetylene can be readily hydrogenated by tristriphenylphosphine halides, Rh(PPh$_3$)$_3$X. The significant characteristic here is

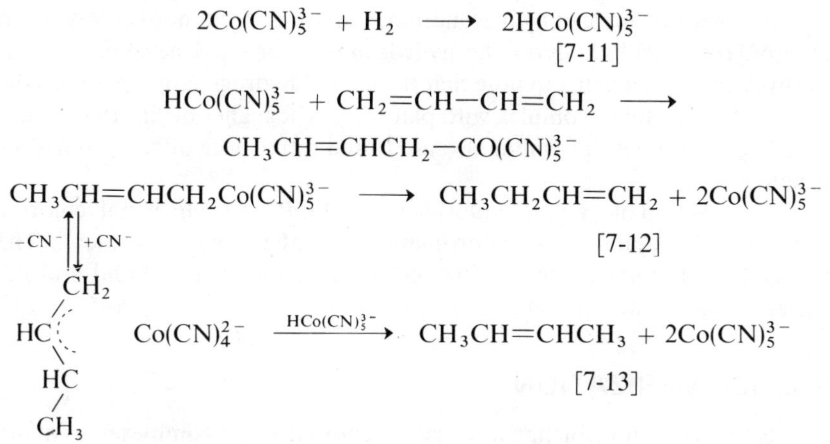

$$2\text{Co(CN)}_5^{3-} + \text{H}_2 \longrightarrow 2\text{HCo(CN)}_5^{3-}$$
$$[7\text{-}11]$$

$$\text{HCo(CN)}_5^{3-} + \text{CH}_2{=}\text{CH}{-}\text{CH}{=}\text{CH}_2 \longrightarrow$$
$$\text{CH}_3\text{CH}{=}\text{CHCH}_2{-}\text{CO(CN)}_5^{3-}$$

$$\text{CH}_3\text{CH}{=}\text{CHCH}_2\text{Co(CN)}_5^{3-} \longrightarrow \text{CH}_3\text{CH}_2\text{CH}{=}\text{CH}_2 + 2\text{Co(CN)}_5^{3-}$$

$-\text{CN}^- \big\Vert +\text{CN}^-$ $[7\text{-}12]$

$$
\begin{matrix}
\text{CH}_2 \\
\text{HC} \\
\; \\
\text{HC} \\
\text{CH}_3
\end{matrix}
\quad \text{Co(CN)}_4^{2-} \xrightarrow{\text{HCo(CN)}_5^{3-}} \text{CH}_3\text{CH}{=}\text{CHCH}_3 + 2\text{Co(CN)}_5^{3-}
$$
$$[7\text{-}13]$$

Fig. 7-3. Co(CN)_5^{3-}-catalyzed hydrogenation of butadiene in an aqueous solution.

the yet unexplained dissociation, in solution, giving reactive bistriphenyl-phosphine derivatives with a coordinate ligand vacancy that Wilkinson describes as "coordinated unsaturation," equation (7-1). The dissociated

$$\text{Rh(Ph}_3)_3\text{Cl} \rightleftharpoons \text{Ph(PPh}_3)_2\text{Cl} + \text{PPh}_3 \qquad (7\text{-}1)$$

derivative readily reacts with carbon monoxide and with olefins to give a *trans* triphenylphosphine olefin complex, equation (7-2). As shown, Co(II) is

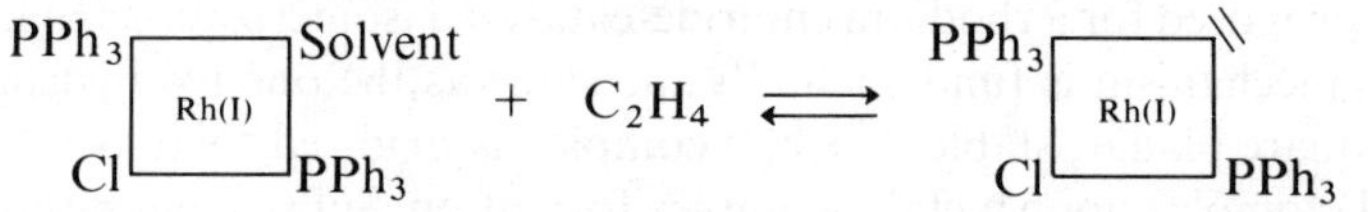

$$
\begin{matrix}
\text{PPh}_3 & \text{Solvent} \\
 & \text{Rh(I)} \\
\text{Cl} & \text{PPh}_3
\end{matrix}
\quad + \quad \text{C}_2\text{H}_4 \rightleftharpoons \quad
\begin{matrix}
\text{PPh}_3 & \\
 & \text{Rh(I)} \\
\text{Cl} & \text{PPh}_3
\end{matrix}
\qquad (7\text{-}2)
$$

oxidized to Co(III) as the intermediate hydride is formed [7-11]. The overall mechanism is supported by the fact that under high CN^- concentrations the hydrogenation predominantly yields 1-butene [7-12] and at low CN^- concentrations 2-butene [7-13] is obtained. The chemical reactivity of this Co(I) complex and related pentacoordinated d^9 complexes resemble that of an organic free radical reactivity. The behavior of Co(CN)_5^{3-} is best understood on the basis of its molecular orbital picture. Six d electrons reside in the t_{2g} orbital, and one electron is located in the eg^* orbital. This latter electron is likened to an organic radical.

Hydrogenation of ethylene by tris(triphenylphosphine) rhodium(I) chloride (Fig. 7-1), is an example of an oxidative addition mechanism. As required, the metal is formally oxidized (Rh(I) to Rh(III)) and a dihydride is formed (tetra- to hexacoordinated configuration).

As a final example of hydrogenation, a mixture of stannous chloride and chloroplatinic acid catalyzes the hydrogenation of ethylene almost quantitatively. It is interesting to note that the ease of hydrogenation parallels the ability of the olefin to complex with platinum. The higher olefins that do not readily complex with platinum also are found to be more difficult to hydrogenate.

The reaction of stannous chloride with a Pt, Pd, Rh, etc. metal–chloride bond usually results in the coordination of $SnCl_3^{(-)}$ with the metal. The $SnCl_3^-$ ligand is known to stabilize both the metal–hydride bond and the metal–olefin bond.

7-2. ISOMERIZATION

Isomerization of olefins have been reported using complexes of many transition metals including iron, cobalt, molybdenum, rhodium, iridium, nickel, palladium, and platinum. In most cases the more thermodynamically favored isomer is formed; thus

Terminal olefin $\longrightarrow$ internal olefin
cis Olefin $\longrightarrow$ *trans*-olefin
Nonconjugated diene $\longrightarrow$ conjugated diene

Labeling studies have shown that migration of the double bonds does not result from successive migration of hydrogen. Evidence for the following type of mechanism, which is probably applicable to other catalysts as well, has been proposed for a rhodium chloride-catalyzed isomerization (Fig. 7-4).

This mechanism is fundamentally the same as the one for hydrogenation. A square planar, stable d^8 Rh(I) complex is oxidized to an octahedral d^6 Rh(III) complex upon metal π-complex formation. Subsequent π-to-σ and σ-to-π rearrangements produce the isomerized product upon reduction of Rh(III) back to Rh(I). It should be noted that during the course of the rearrangement (structures [7-16] and [7-17] a *cis* ligand orientation effect is operative between the propylene molecule and the hydrogen atom.

At this point we shall examine more closely the substitution process represented by structures [7-15] and [7-16]. The principles to be presented are equally valid for similar six and four coordination species substitution reactions involved in hydrogenation, dimerization, polymerization, and oxidation—among others.

For a substitution reaction involving a six-coordinated species we can envision two possible transition states. One involves an S_N2 mechanism (substitution, nucleophilic, and bimolecular) where a six-coordinate complex accepts the ligand to be added. This produces a seven-coordinate system as the intermediate that then releases one of the substituted ligands, thus returning to a new six-coordinate system. The second possible transition state involves an S_N2 mechanism

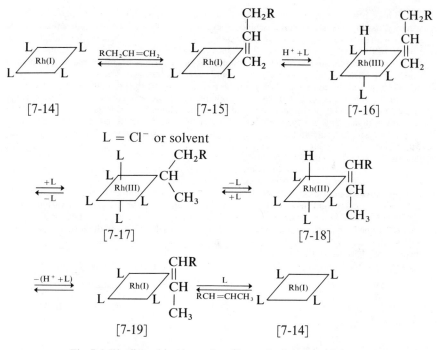

Fig. 7-4. Rhodium chloride-catalyzed isomerization of an olefin.

(substitution, nucleophilic, and unimolecular) and a five-coordinate square pyramidal, inter-mediate is formed upon elimination of a ligand prior to accommodation of the approaching substituent. Either route will produce a new six-coordinate complex (Fig. 7-5).

In order to proceed through a seven-coordinate transition state the energy of activation is low (species will be labile) if a vacant d orbital can be utilized. However, if all the d orbitals are filled (d^{10}) one electron must vacate a d orbital and a higher activation energy is required for the transition state, resulting in a slower reaction rate. The S_N1 dissociation process to a five-coordinate transition state is favored in species in which some of the anti-bonding orbitals are occupied. If all d orbitals are filled d^{10} two things can occur in forming a heptacoordinated metal: (1) the ligand may add electrons to outer available orbitals or (2) $2d$ electrons must "vacate" d orbitals or must be promoted to higher energy states.

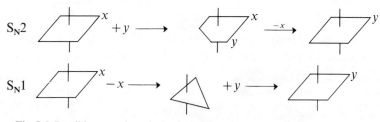

Fig. 7-5. Possible routes for substitution of ligands in a six-coordinate species.

Similar observations can be made for coordination number four species. Since tetrahedral species (possibly) possess vacant d orbitals, they are generally more labile, while square planar species with no vacant d orbitals are more stable. Thus the activation energy for the latter case is high. In the present example (Fig. 7-4, steps [I] and [II]) the S_N1 mechanism seems favored. Rh(I) is a d^8 ion in a square planar dsp^2 configuration. Thus it is assumed that all the metal d orbitals are filled and therefore would not readily accept additional electrons as required by an S_N1 mechanism. The energy level diagrams in Fig. 7-6 support the predicted configurations.

Furthermore, it is suggested that for square planar species two solvent molecules probably occupy perpendicular positions above and below the plane. Thus the mechanism for such a complex actually may involve a six-coordinate species with a distorted octahedral structure (Fig. 7-7). Again favoring an S_N1 mechanism it is postulated that substitution involving ligands with weak *trans* directing properties go via path 1 and those involving strong *trans* directing ligands go via path 2 (Fig. 7-7).

The crystal field stabilization energy calculated for the various geometric forms can be equated to the likelihood of distorting or even converting, for example, a five-coordinate species to a six- or seven-coordinate species. Thus, if the crystal field stabilization energy for the transition state is greater than that for the original six coordinate specie, then the activation energy for the process will be lower and the species will be labile. If the reverse is true, then the activation energy will be high and the specie will be inert. For strong field ligands the degree of inertness varies: $d^8 \sim d^5 < d^4 < d^3 < d^6$. For a weak field only d^3 and d^8 should be inert. The experimental observations fit the calculated data reasonably well.

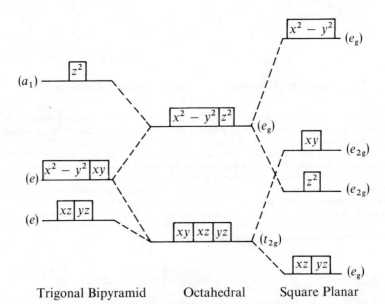

$$\text{Trigonal Bipyramid} \qquad \text{Octahedral} \qquad \text{Square Planar}$$

Fig. 7-6. Energy level diagram depicting orbital splitting in complexes of various symmetries. Only the orbitals involved in accommodating the d electrons of the metal are depicted. Note that the octahedral configuration with only three stable (t_{2g}) orbitals is most favorable for accommodation of six d electrons, whereas the two configurations of lower coordination number, each with four stable orbitals, are more favorable for the accommodation of seven or eight d electrons. The orbital symmetries are described by the symbols $xz, yz, x^2 - y^2, \ldots$ as well as by the group theoretical designations $a, e, t_{2g}. \ldots$

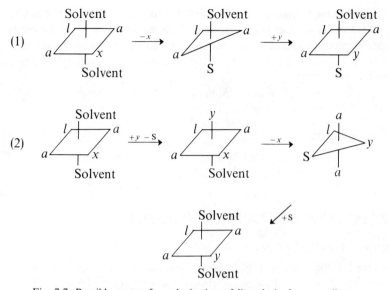

Fig. 7-7. Possible routes for substitution of ligands in four-coordinate square planar species.

$$RCH_2-CH=CH_2 \rightleftharpoons RCH \overset{H}{\underset{\underset{H}{\overset{|}{M}}}{\overset{C}{\diagup \diagdown}}} CH_2 \rightleftharpoons RCH=CH-CH_3$$

Fig. 7-8. Alternate isomerization mechanism.

It can be shown experimentally, that four coordinate complexes of palladium in general are more labile than platinum but more inert than nickel complexes. This reflects the greater crystal field stabilization energy of $4d$ and $5d$ elements, that is, Δ increases with increasing quantum number.

Isomerization also may proceed via the rearrangement of an intermediate π-allyl hydride, as shown in the scheme in Fig. 7-8.

1,3-Cyclo-octadiene can be converted to the less thermodynamically stable 1,5-isomer using a rhodium(III) chloride $RhCl_3 \cdot 3H_2O$ catalyst, equation (7-3).

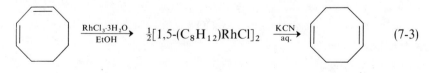

$$\xrightarrow[\text{EtOH}]{RhCl_3 \cdot 3H_2O} \tfrac{1}{2}[1,5\text{-}(C_8H_{12})RhCl]_2 \xrightarrow[\text{aq.}]{KCN} \tag{7-3}$$

The reverse isomerization of 1,5 to 1,3-cyclo-octadiene has been reported with the use of iron pentacarbonyl. This isomerization can be predicted from thermodynamic considerations. A relatively unstable diene iron(III) carbonyl intermediate (non planarity of 1,3-cyclo-octadiene) is probably operative (Fig. 7-9).

Isomerization of a nonconjugated olefin to a conjugated olefin by $Fe(CO)_5$ is shown in equation (7-4).

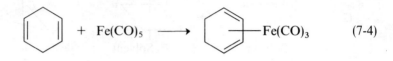

$$+ \quad Fe(CO)_5 \quad \longrightarrow \quad \text{—}Fe(CO)_3 \qquad (7\text{-}4)$$

7-3. POLYMERIZATION OF OLEFINS

Ziegler–Natta Process

One of the triumphs of modern chemistry is the polymerization of mono-olefins such as ethylene and propylene. During the 1950's ethylene was polymerized using a Ziegler–Natta catalyst—a mixture of transition metal halides (titanium halides are particularly active) and trialkylaluminum (triethylaluminum is commonly used). The technological importance to, and wide application in, the plastic and fiber industries stimulated extensive studies by industrial and academic researchers in determining the scope and the mechanism of this polymerization process. A large number of patents and publications have been reported. The use of trialkylaluminum in the process stimulated a broad investigation of the use of organometallic compounds in general. As the mechanism of the polymerization and the nature of the catalyst became clearer, it has been determined that the Ziegler–Natta process definitely involves a metal π-complex intermediate. A plausible mechanism for the polymerization can be formulated by applying typical organometallic and coordination reactions. The first step involves the "alkylation" of

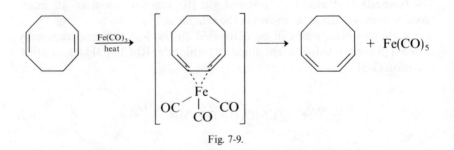

Fig. 7-9.

titanium ions. The R group from AlR_3 was found to be located at the end of the polymer chain formed by the catalyst, a fact that has been determined from tracer studies (Fig. 7-10).

When the alkylation step was recognized as essential, various alkyltitanium compounds were employed as modified Ziegler–Natta catalysts. It is reasonably postulated that ethylene coordinates to the titanium at positions *cis* to the R ligand. This *cis* ligand orientation effect on the metal ion is essential to polymerization.

The type of cleavage between the titanium metal and the R carbon largely depends on the nature of the anionic ligands and the solvents, among other factors. It is reasonable that homolytic cleavage usually occurs and a bond is formed between the R carbon and the ethylene–carbon, while a σ-bond is formed between the ethylene carbon and the titanium metal ion. This creates a vacant site on the metal ion *cis* to the newly extended organic group. A new ethylene molecule can be inserted between the titanium metal ion and the organic group as described above. This process can be stereospecifically controlled by the nature of the ligands and solvents or by the addition of other components such as amines. The chain length or molecular weight of the polyethylene can be adjusted by the addition of hydrogen or carbon monoxide, among others.

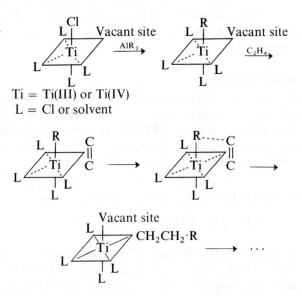

Fig. 7-10. Mechanism of the Ziegler–Natta process for polymerization of ethylene.

Stereoregular Polymerization of Propylene

Propylene is polymerized by various protonic or Lewis acid catalysts to a viscous, oily polymer. The use of a Ziegler–Natta catalyst, especially the combination of solid $TiCl_3$–$AlEt_3$, produced a remarkable new crystalline polymer that has a melting point as high as 165°C. The polymer exhibits a high degree of stereoregularity [7-15]

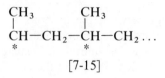

[7-15]

A regular sequence involving a *ddd*... or *lll*... configuration with respect to the asymmetric carbon in the polymer chain is called an isotactic polymer, while a *dl dl*... arrangement is referred to as a syndiotactic polymer. Isotactic polypropylene has been extensively used as a textile material or as a plastic.

A number of different mechanisms have been proposed for isotactic polymerization (stereoregular polymerization). Among these a mechanism found wide acceptance utilizing the stereo selective adsorption (π-complex formation) of propylene over a vacant site on the surface of $TiCl_3$ produced by the action of $AlEt_3$. Adsorption of propylene in a stereo selective manner and subsequent insertion to the alkyl group, as mentioned above, gives isotactic polymer.

Co-Oligomerization

Recently a codimerization of ethylene and butadiene using various organometallic catalyst systems to give *cis*-1,4 hexadiene was found, equation (7-5). Hexadiene is a potential third component (or termonomer) of ethylene-propylene-diene-rubber [sometimes called ethylene propylene-terpolymer, EPT or EPDM (according to the ASTM)].

$$CH_2=CH_2 + CH_2=CH-CH=CH_2 \longrightarrow$$

$$CH_2=CH-CH_2-CH=CH-CHO \tag{7-5}$$

Among many catalyst systems, Ziegler combinations, for example, $Fe(acac)_3$–$AlEt_3$, CoX_2–$AlEt_3$-diphosphine and $NiX_2(PR_3)_2$–$AlEt_3$, are described in various patents. A pure π-complex, $Fe(0)(C_8H_8)_2$, also can be used. Rhodium chloride ($RhCl_3 \cdot nH_2O$) in ethanol was found effective. A π-allyl intermediate was proposed for this catalyst, equation (7-6).

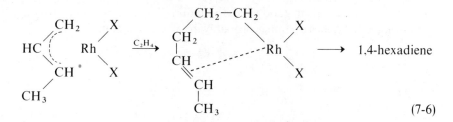

$$\text{(7-6)}$$

The cyclo-co-oligomerization of 2 moles of butadiene and 1 mole of ethylene also was catalyzed by a Ziegler combination, $Ni(acac)_2$–Et_2AlOEt, to give *cis-trans*-1,5-cyclodecadiene, equation (7-7). A related linear co-

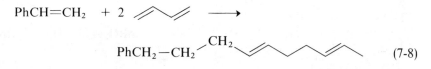

$$\text{(7-7)}$$

oligomerization has been reported using styrene in place of ethylene, equation (7-8). The product is a starting material for a "soft-type" synthetic detergent.

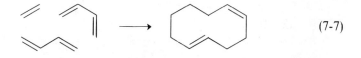

$$\text{(7-8)}$$

Linear Oligomerization

Linear dimerization of ethylene is catalyzed, among others, by rhodium chloride ($RhCl_3 \cdot 3H_2O$) in ethanol. The proposed mechanism for this dimerization is summarized in Fig. 7-11. Various Ziegler combinations using Ni, Co, and Ti compounds also are effective.

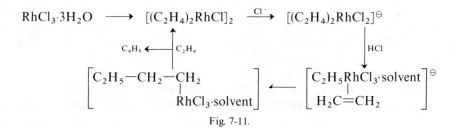

Fig. 7-11.

Dimerization of butadiene with organometallic cobalt π-complexes, such as $(\pi\text{-}C_3H_5)_3Co$, $(1,5\text{-cycle } C_8H_{12})_2Co$, or $[Co(CO)_2(C_4H_6)]_2$, gives branched chain dimers in good yields, equation (7-9). An intermediate complex isolated from the reaction is probably involved, equation (7-10).

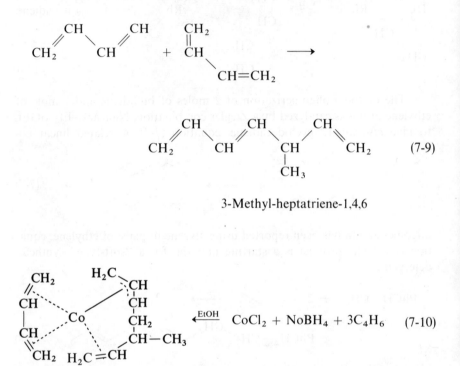

(7-9)

3-Methyl-heptatriene-1,4,6

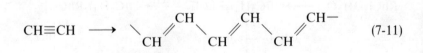

$$\xleftarrow{\text{EtOH}} CoCl_2 + NoBH_4 + 3C_4H_6 \qquad (7\text{-}10)$$

Polymerization of Multiply Unsaturated Compounds

Acetylene has been polymerized by various organo-nickel catalysts or $TiX_3\text{–}AlR_3$ (a typical Ziegler catalyst) to a black, insoluble powder having conjugated double bonds, equation (7-11). Alkylacetylenes give a similar

$$CH{\equiv}CH \longrightarrow \qquad\qquad\qquad (7\text{-}11)$$

polymer and in some cases give linear dimer, trimers, and tetramers. A π-complex intermediate that is in equilibrium with a σ-complex has been proposed, equation (7-12).

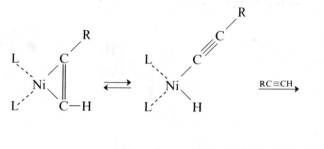

$$R-C\equiv C-CH-CH-R \qquad (7\text{-}12)$$

A polymer of allene can be prepared with typical Ziegler catalysts or more easily with various π-complex catalysts, for example, $(\pi\text{-}C_3H_5)_2Ni$, $(1,5\text{-}C_8H_{12})_2Ni$, and $[(\pi\text{-}C_3H_5)NiBr]_2$, equation (7-13).

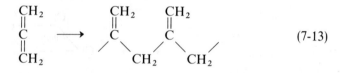

$$(7\text{-}13)$$

7-4. OXIDATION OF OLEFINS (WACKER PROCESS)

The oxidation of ethylene exclusively to acetaldehyde (and other straight-chained olefins to ketones) is achieved by the catalytic reaction of ethylene in an aqueous solution by palladium(II) chloride or by oxygen in the presence of palladium(II) chloride, copper(II) chloride, or iron(III) chloride. Generally, the oxidation of olefins by other metal ions, such as Hg(II), Th(III), and Pb(IV) yields glycol derivatives as well as carbonyl products. The mechanism for the oxidation is plausibly accommodated in the scheme shown in Fig. 7-12.

Evidence for a hydride internal migration also was presented in the reaction of $(C_2H_4PdCl_2)_2$ with deuterium oxide, which gave acetaldehyde free from deuterium. This indicates that vinyl alcohol was not a reaction intermediate. The lack of exchange is also consistent with a direct 1,2-hydride shift.

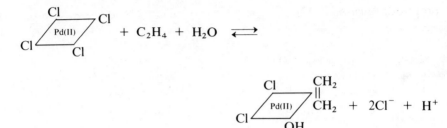

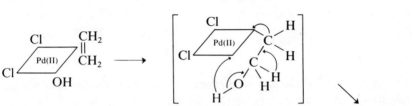

$$Pd(0) + H^+ + 3Cl^- + CH_3CHO$$

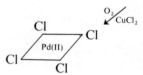

Fig. 7-12. Possible mechanism for ethylene oxidation (Wacker process).

7-5. ADDITION OF CARBON MONOXIDE

Oxo Reaction

The oxo reaction is probably the best-known addition reaction of carbon monoxide to olefins (or acetylenes). The oxo process has been developed extensively in industry to produce primary alcohols by the reduction of the

$$RCH{=}CH_2 + H_2 + CO \xrightarrow{\;H^*Co(CO)_4\;} RCH_2CH_2CHO$$

$$^*[Co_2(CO)_8 + H_2 + CO \longrightarrow HCo(CO)_4] \qquad\qquad (7\text{-}14)$$

aldehydes thus formed. Olefinic hydrocarbons are reacted with carbon monoxide and hydrogen in the presence of $HCo(CO)_4$ as a catalyst to yield

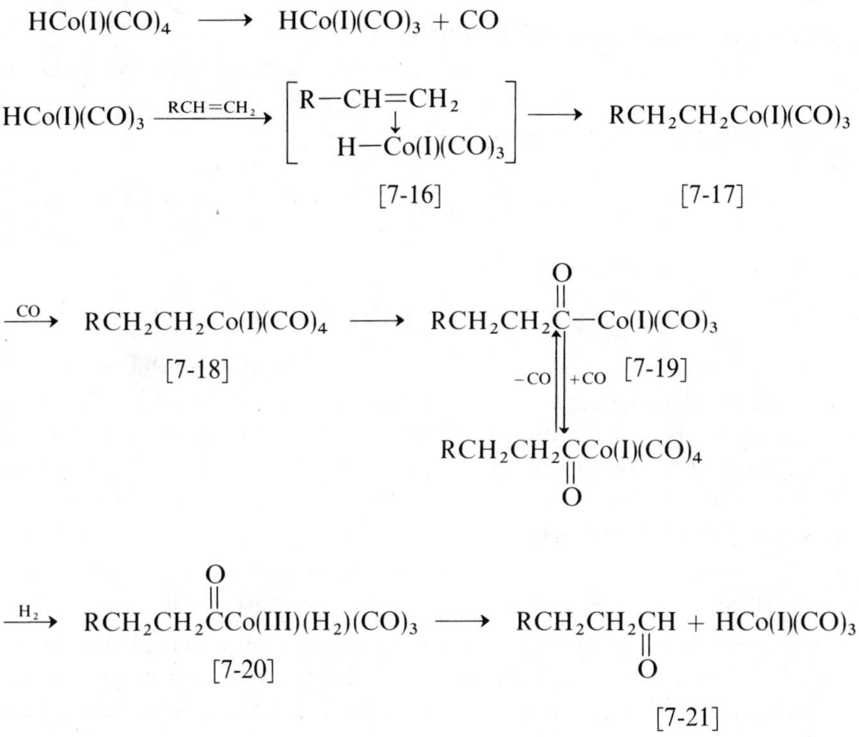

$$HCo(I)(CO)_4 \longrightarrow HCo(I)(CO)_3 + CO$$

$$HCo(I)(CO)_3 \xrightarrow{RCH=CH_2} \begin{bmatrix} R-CH=CH_2 \\ \downarrow \\ H-Co(I)(CO)_3 \end{bmatrix} \longrightarrow RCH_2CH_2Co(I)(CO)_3$$

$$[7\text{-}16] \qquad\qquad\qquad [7\text{-}17]$$

$$\xrightarrow{CO} RCH_2CH_2Co(I)(CO)_4 \longrightarrow RCH_2CH_2\overset{\displaystyle O}{\overset{\displaystyle \|}{C}}-Co(I)(CO)_3$$

$$[7\text{-}18] \qquad\qquad\qquad\qquad -CO \;\Big\|\, +CO \quad [7\text{-}19]$$

$$RCH_2CH_2\underset{\displaystyle \underset{O}{\|}}{C}Co(I)(CO)_4$$

$$\xrightarrow{H_2} RCH_2CH_2\overset{\displaystyle O}{\overset{\displaystyle \|}{C}}Co(III)(H_2)(CO)_3 \longrightarrow RCH_2CH_2\underset{\displaystyle \underset{O}{\|}}{CH} + HCo(I)(CO)_3$$

$$[7\text{-}20] \qquad\qquad\qquad\qquad\qquad [7\text{-}21]$$

Fig. 7-13. Mechanism for the oxo reaction.

aldehydes, equation (7-14). This catalyst is formed from dicobalt octacarbonyl and molecular hydrogen. A plausible mechanism (Fig. 7-13) was formulated after extensive investigation.

It has been proposed that the formation of a π-complex intermediate [7-16] is essential to the reaction. A hydride ion transfer from a Co(I) species [although $HCo(CO)_4$ is acidic, it can be formally regarded as $H^{(-)}$] to the β-carbon atom causes a $\pi-\sigma$ rearrangement that yields [7-17]. Carbon monoxide is then added to the Co(I) ion. One of the four carbonyls on Co(I) [7-18] is inserted between Co(I) and the α-carbon atom of the hydrocarbon group. Subsequent oxidative addition of a hydrogen molecule to give two hydrogen atoms on Co(I) in [7-19] results in the formation of the stable Co(III) (d^6) octahedral configuration [7-20]. Cleavage of [7-20] into two molecules produces the aldehyde [7-21] and results in the concurrent reduction of the formal Co(III) to Co(I) with the formation of $HCo(I)(CO)_3$.

Carbonylation (Reppe Reaction)

Nickel carbonyl reacts stoichiometrically with acetylene to yield an acrylic ester, equations (7-15) and (7-16). A plausible mechanism via a π-complex intermediate is proposed (Fig. 7-14).

$$HC\equiv CH + CO + ROH \xrightarrow[\text{high pressure}]{\text{Ni catalyst}} CH_2=CHCOOR \qquad (7\text{-}15)$$

$$4HC\equiv CH + 4ROH + Ni(CO)_4 + 2HCl \longrightarrow$$

$$4CH_2=CHCOOR + H_2 + NiCl_2 \qquad (7\text{-}16)$$

It is known that in the stoichiometric carbonylation of $PhC\equiv CH$ and $PhC\equiv CPh$ the hydrogen atom and the later group are added via a *cis* addition. By this mechanism the essential role of a π-complex intermediate [7-22] in the stereospecific *cis* addition is clarified. The π-allyl nickel(II) halide complex $(\pi\text{-}C_3H_5NiX)_2$ reacts with carbon monoxide to give various products with different coreactants.

A mechanism for the formation of β-unsaturated acyl halides from $(\pi\text{-}C_3H_5NiX)_2$ (X = Br or I) and CO has been suggested, equation (7-17).

Attempts to isolate [7-24] however, have been unsuccessful. The initial step is supposedly the formation of a halonickel dicarbonyl hydride [$HNi(CO)_2X$], which adds to the unsaturated compound with subsequent

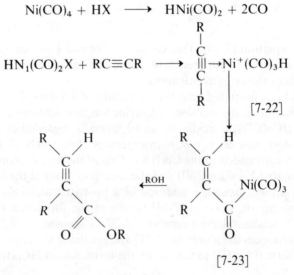

Fig. 7-14.

$$(\pi\text{-}C_3H_5NiX)_2 + 6CO \longrightarrow 2CH_2{=}CH{-}CH_2{-}\overset{\displaystyle CO}{\underset{\displaystyle CO}{\overset{|}{\underset{|}{C}}ONiX}} \xrightarrow{\text{4CO}}$$

$$[7\text{-}24]$$

$$2CH_2{=}CH{-}CH_2{-}COX + 2Ni(CO)_4 \qquad (7\text{-}17)$$

carbon monoxide absorption [7-24] followed by the elimination of the acyl halide. The formation of the alkyl halide and its direct reaction with nickel carbonyl to give [7-24] also is a possible first step. The allylic complex is then carbonylated to give the corresponding acyl halides. It is believed that the driving force for the reaction is the coordination of carbon monoxide achieved either by halide bridge splitting or by the expansion of the co-ordination shell of the metal.

A π-allyl derivative of palladium(II) chloride is first formed, to which carbon monoxide can be added (Fig. 7-15A). Similarly, an olefin is converted to the acid chloride in the presence of palladium metal, hydrochloric acid, and carbon monoxide (Fig. 7-15B).

In asymmetric olefins the carbon monoxide attack occurs at the most substituted carbon, although olefins higher than pentene behave differently. Carbonylation of π-allylic complexes of palladium also takes place at the carbon with the least number of hydrogens and upon hydrolysis gives unsaturated acids.

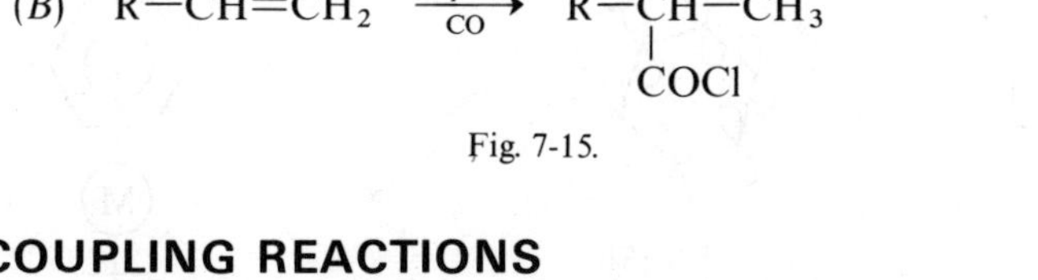

$$(A) \quad \begin{array}{c} CH_2{=}CHCH_2OH \\ \text{or } CH_2{=}CH{\cdot}CH_2Cl \end{array} \xrightarrow{\text{PdCl}_2} (\pi\text{-}C_3H_5PdCl)_2$$

$$\Big\downarrow \begin{array}{l} \text{2CO} \\ \text{100 Kg/cm}^2 \\ \text{70°, 5 hr} \end{array}$$

$$CH_2{=}CHCH_2COCl$$

$$(B) \quad R{-}CH{=}CH_2 \xrightarrow[\text{CO}]{\text{Pd, HCl}} R{-}\underset{\displaystyle COCl}{\overset{|}{CH}}{-}CH_3$$

Fig. 7-15.

7-6. COUPLING REACTIONS

Although many coupling reactions of olefins have been described, the most familiar one is an aryl coupling reaction.

The reaction of phenylmagnesium bromide and chromium(III) chloride in ether, initiated at room temperature, results in a quantitative yield of

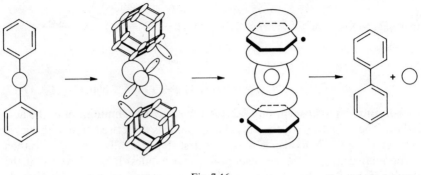

Fig. 7-16.

biphenyl. A plausible mechanism for this stoichiometric phenyl coupling reaction was proposed by applying a metal π-complex mechanism, as shown in Fig. 7-16. The first step may involve the formation of a σ-bonded triphenyl chromium trietherate, an octahedral chromium d^3 complex. However, under

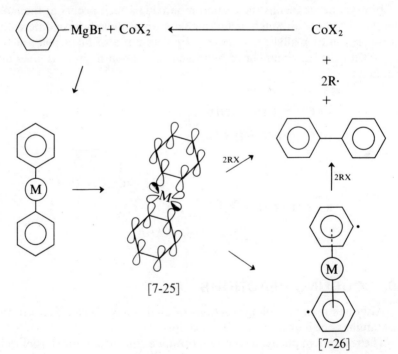

Fig. 7-17.

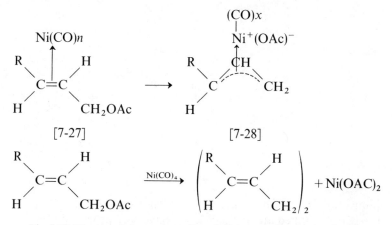

Fig. 7-18. Deacetoxylation and coupling of allyl acetate by nickel carbonyl.

the reaction conditions the weakly stabilizing ether ligands are easily removed, resulting in the formation of an unstable σ-bonded triphenyl chromium(III). Homolytic cleavage of the metal–carbon bonds results in the formation of radical arenes (phenyl radicals) and a very active zero-valent chromium. Overlap of arene π-electrons with the vacant chromium d orbitals would give diradical metal π-complex intermediate. Some degree of stability is imparted by the filling of all the valence orbitals of this d^3 metal by 12 ligand electrons. Under the reaction conditions the decomposition of the intermediate gives biphenyl. It should be noted that under milder conditions, 0–5°C, mixed arene chromium π-complexes are isolated.

The catalytic aryl coupling reaction called the Kharasch–Grignard reaction consists of the rapid formation of biphenyl by the addition of an organic halide to a mixture of phenyl magnesium bromide and a trace amount of a transition metal halide. A π-complex mechanism from this reaction has been proposed, as shown in Fig. 7-17. According to this mechanism an organic halide most likely attacks the metal atom in [7-25] or [7-26].*

A different type of catalytic coupling reaction, but also involving a metal π-complex intermediate, is exemplified by the deacetoxylation and

* The stoichiometric reaction of cis- or trans-styryl, magnesium bromide, and chromium(III) chloride or cobalt(II) chloride in tetrahydrofuran at 30°C yields 1,2,3,4-cis, cis-tetraphenylbutadiene. Using palladium(II) chloride or nickel bromide, a different coupled product, 1,2,3,4-trans, trans-tetraphenyl butadiene, can be obtained. The mechanism of this stereospecific coupling reaction also involves metal π-complex formation. Those transition metal halides usually exhibiting octahedral d^2sp^3 orbital hybridization give the cis-cis product, and those exhibiting a square planar dsp^2 or tetrahedral sp^3 orbital hybridization give the trans-trans product. A more detailed mechanismistic study is required.

$$CH_3C{\equiv}CCH_3 + Ph_3Cr{\cdot}3THF \longrightarrow$$

[7-29]

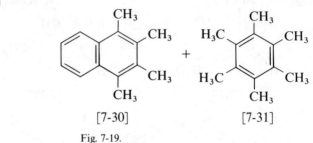

[7-30]   [7-31]

Fig. 7-19.

coupling of allyl acetate by nickel carbonyl. The formation of [7-27] and [7-28] (Fig. 7-18) as intermediates in the reaction have been proposed.

7-7. CYCLIZATION

σ-Bonded aryl and alkyl transition metals catalize the cyclization of disubstituted acetylenes. For instance, triphenylchromium tritetrahydrofuranate [7-29] reacts with butyne-2 to form hexamethylbenzene [7-30] and 1,2,3,4-tetramethyl naphthalene [7-31] (Fig. 7-19). Phenyl groups are incorporated into the reaction product. It is reasonable to speculate that dehydrogenation at the β-carbon atom was catalyzed by a chromium atom. A plausible mechanism for this cyclization reaction is shown in Fig. 7-20.

Butyne-2 replaces the coordinating tetrahydrofuran ligands in a stepwise fashion ([7-33] to [7-34]), producing a tetramethylcyclobutadiene π-complex. Butyne-2 is added to tetramethyl cyclobutadiene in a Diels–Alder-type addition giving [7-36], which can be envisioned as a hexamethyldicyclohexadiene 2,0,2 (Dewar benzene) π-complex. Further reaction of butyne-2 with [7-36] gives bis-(hexamethyl benzene) chromium (isolated). Upon hydrolysis of the reaction mixture, hexamethyl benzene [7-37] and benzene are obtained. If complex [7-34] is partially decomposed, dehydrogenation of one of the phenyl groups by a chromium metal atom may produce a benzyne species that may attack a cyclobutadiene molecule, producing [7-35]. Similar incorporation of organic groups are given by reactions (7-18), (7-19), and (7-20).

In the cyclization of diphenyl acetylene, methyl and ethyl groups are incorporated into the products. The greater dehydrogenating power of chromium over nickel is demonstrated by comparing reaction (7-17) with (7-18).

Possibly because of the transitory stability of a π-complex intermediate, 1,2,4-tri,-*t*-butylbenzene was prepared utilizing the π-complex; many

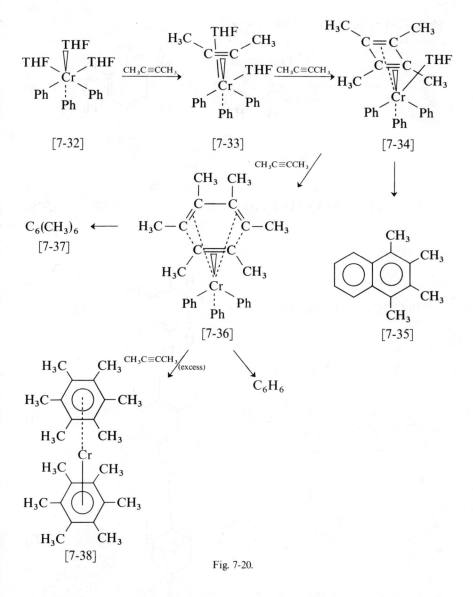

[7-32]   [7-33]   [7-34]

$C_6(CH_3)_6$

[7-37]

[7-36]   [7-35]

[7-38]

Fig. 7-20.

attempts by a more direct route proved unsuccessful because of severe steric hindrance. The π-complex intermediate, structure [7-40] in Fig. 7-21 can be formed when the steric compression is not prohibitively severe. Other examples of the preparation of hindered compounds by a cyclization reaction are given in equations (7-21) and (7-22).

$PhC{\equiv}CPh + (CH_3)_3Cr{\cdot}3THF \longrightarrow$

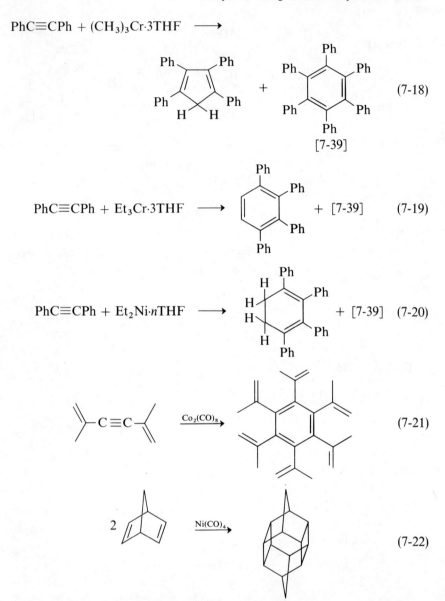

(7-18)

[7-39]

$PhC{\equiv}CPh + Et_3Cr{\cdot}3THF \longrightarrow \qquad + [7\text{-}39]$ (7-19)

$PhC{\equiv}CPh + Et_2Ni{\cdot}nTHF \longrightarrow \qquad + [7\text{-}39]$ (7-20)

$\xrightarrow{Co_2(CO)_8}$ (7-21)

$2 \qquad \xrightarrow{Ni(CO)_4} \qquad$ (7-22)

The widely known Repp's synthesis of cyclo-octatetraene by tetra-merization of acetylene undoubtedly involves a metal π-complex inter-mediate. However, a clear-cut mechanism for this reaction has not yet been advanced. The formation of a cyclobutadiene ring by cyclodimerization of

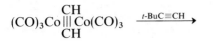

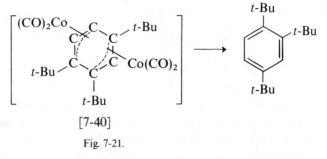

[7-40]

Fig. 7-21.

acetylenic compounds has been reported. One of the classic examples of this reaction is shown in Fig. 7-22. As indicated, the reaction mechanism is best described by a sequence involving the substitution of a square planar palladium derivative. Diphenylacetylene is also cyclized by iron pentacarbonyl (Fig. 7-23).

Decomposition of structure [7-41] in Fig. 7-22 and structure [7-42] in Fig. 7-22 by triphenylphosphine or by heating gives octaphenylcyclooctatetraene.

$$PhC\equiv CPh + (C_6H_5CN)_2PdCl_2 \xrightarrow{EtOH-ChCl_3}$$

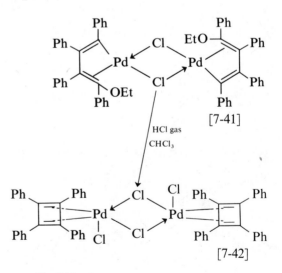

Fig. 7-22.

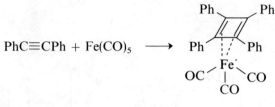

$$PhC{\equiv}CPh + Fe(CO)_5 \longrightarrow$$

Fig. 7-23.

Cyclodimerization or trimerization of straight chain mono-olefins using metal complexes have not been reported. However, norbornadiene was found to give a cyclodimerization product, equation (7-23).

(7-23)

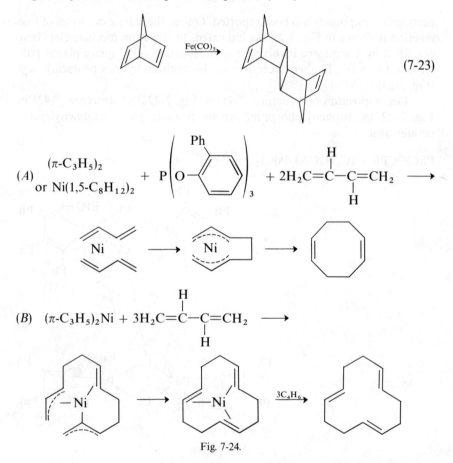

Fig. 7-24.

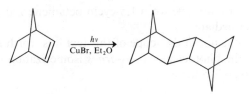

Fig. 7-25. Stereo selective photodimerization of
norbornene-catalyzed by CuBr.

1,3-Butadiene has been successfully cyclodi- and trimerized to cyclo-octa-
diene and cyclododecatriene by nickel catalysts such as Ni-(0)-[P(C$_6$H$_5$)$_3$]$_4$,
Ni-(0)-[AS(C$_6$H$_5$)$_3$]$_4$, and (π-allyl)$_2$Ni. The mechanisms for cyclodi- and
trimerization of 1,3-butadiene are shown in Fig. 7.24A and B.

7-8. PHOTOCHEMICAL REACTIONS

Organic photochemistry has had an impact in almost every area of
chemistry. Only limited data are available in the metal π-complex field; the
subject is nonetheless potentially stimulating. Inter- and intramolecular
photoreactions apparently involve preformed metal olefin π-complexes.

Intermolecular Photoreactions

Ultraviolet irradiation of ether solutions of norbornene in the presence
of copper(I) halides causes the stereo selective formation (97%) of an *exo*,
trans, *exo* dimer (Fig. 7-25). Coordination of the copper atom with norbornene is suggested.

Complex formation between norbornene and CuX appears essential.
The complex subsequently catalyzes the dimerization.

Intramolecular Photodimerizations

Mercury sensitized photolysis of 1,5-cyclo-octadiene in the vapor phase
gives cyclo[5.1.0]octene-3 (Fig. 7-26).

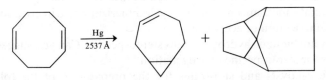

Fig. 7-26. Mercury sensitized photolysis of 1,5-cyclo-octadiene.

Complex formation between 1,5-cyclo-octadiene and mercury is postulated as an intermediate.

Although it is not in the scope of this text it is worth noting that ferrocene catalyzes the photochemical *cis–trans* isomerization of piperylene and the dimerization of isoprene.

7-9. EXERCISES

7-1. Illustrate typical examples of the mode of reaction of a hydrogen molecule with metal complexes.

7-2. List the catalytically active metal ions in homogeneous hydrogenation of olefins.

7-3. Show the direction of isomerization by $Fe(CO)_5$ catalyst.

7-4. What combination of metal compounds are catalytically effective for isotatic polypropylene synthesis?

7-5. What kind of reactions are catalyzed by π-allyl complexes?

7-6. What is Kharasch's catalytic aryl coupling reaction?

7-7. What type of catalyst is effective for alkyne cyclization?

7-8. What products are expected from the hydroformylation of α-olefin?

7-9. What catalyst is usually used for the synthesis of cyclo-octatetraene from acetylene?

7-10. What metal π-complexes are active for a *t.t.t.*-cyclododecatriene synthesis by trimerization of butadiene?

7-11. Formulate the sequence of reactions in the "Wacker" process.

7-12. Write as many examples of co-oligomerization as possible.

7-13. Describe *cis* orientation effects using an example and give its significance of this effect in catalysis.

7-14. Explain σ–π and π–σ rearrangements. Can you classify the processes involved in their rearrangements?

7-15. Answer the following questions about the Ziegler–Natta catalyst.
   a. What is the composition?
   b. Which olefins are polymerized by this catalyst?
   c. What modifications can be suggested for this catalyst?
   d. What is the role of the catalyst?

7-16. Discuss some basic principles of homogeneous catalysis.

7-17. Postulate mechanisms for the cyclodimerization, trimerization, and tetramerization of acetylenes.

7-18. Why are biologic hydrogenations stereospecific? Can you suggest catalysts for stereospecific hydrogenation?

7-19. Give catalysts and substrates for the preparation of the following compounds:

a. Hexamethylbenzene   f. *cis*-Polybutadiene
b. Benzene   g. Acetaldehyde
c. Cyclo-octatetraene   h. Vinylacetate
d. Ethylene-propylene copolymer i. Hexaphenylbenzene
e. 1,2,3,4-Tetraphenylnaphthalene j. Octaphenylcyclo-octatetraene

7-10. BIBLIOGRAPHY

F. Basolo and R. Johnson, "Coordination Chemistry," W. A. Benjamin, New York (1964).

F. Basolo and R. G. Pearson, "Mechanism of Inorganic Reactions," John Wiley & Sons, New York (1963).

C. W. Bird, "Transition Metal Intermediates in Organic Synthesis," Logos Press, London (1967).

J. Halpern, *Chem. Eng. News* 69–75, (October 31, 1966).

C. H. Langford and H. B. Gray, "Ligand Substitution Processes," W. A. Benjamin, New York (1965).

L. Reich and A. Schindler, "Polymerization by Organometallic Compounds," Interscience Publishers, New York (1966).

Index